ARE WE AWAKENED YET?

ARE WE AWAKENED YET?

How to Save Our Earth

ILKA HANDSHAW

Are We Awakened Yet?

Published by:
Professional Woman Publishing
www.pwnbooks.com

ISBN: 978-0-9977775-7-4

Dedication

This is to all truth-seekers who love planet Earth – our home – and want to preserve it; to all tree-huggers, nature-lovers and volunteers, keeping it clean and beautiful; to everyone doing something to reverse the climate change and weather-disaster patterns, disrupting normal life for us all over the world.

To all of you already awakened to the TRUTH of all of this – preserving life on Earth, nurturing our "home", and doing your Best to awaken others to do the same: with all my heart and respect to you: thank you and – keep up the good work!

This is the ONLY home we have; we don't have a choice!

To my family, children and grandchildren who will live in a better, cleaner and sustained world because of all of us – awakened beings of our world now chose to inspire, awaken and shift more and more minds to realize what is MOST important and do exactly that – fulfill our MAIN purpose in life: to save and sustain our planet …

Because nothing else would be possible or even available if we didn't.

Ilka Handshaw

Contents

Introduction

"The Universe is a stage; a human is an actor;
he comes, sees and ... defeated he goes..."

Are you still on "your stage" or almost ready to "go"/"go home"?

Or – you might not have arrived yet...?

Wherever you are, one thing I know for sure is – you want to be happy and joyful, because that is our purpose of being "here". You want to live your life to the fullest and to be a life of great quality, not interrupted by a tornado or a hurricane or an earthquake, especially at places where they shouldn't be...

We HAVE problems with our climate, weather, pollution in our world – many of those are obvious as you see the pollution in the water, air, and soil. When it comes to the weather, it doesn't just rain anymore – IT POURS!!! And sometimes and at some places – for days and weeks; sometimes it snows ... for days and weeks and extreme temperatures are not "strange facts" NOT anymore ... And dangerous facts are showing how global temperatures are rising, Arctic ice is melting and sea levels are rising; carbon dioxide levels are the highest ... ever ... and the ozone layer is dissipating.

No one is happy to hear about all this. And – some people, actually are so scared they struggle with anxiety, are depressed, some even suicidal, because they don't see any hope.

But – if we look at *it* from a different "angle": my dear friends, all Humans of our beautiful planet Earth, we have been soooo BLESSED for soooooo many years to live in such a magical , wonderful HOME!!!

There is NO any other place like it in the whole Universe that we know of. Look around – you will see beauty all year long: green trees and plants and all different colored flowers, and blue waters of the seas and oceans and lakes around the world with beautiful beaches that we LOVE. And – perhaps looking forward to enjoying at Summer time (or if you're lucky – you might enjoy these all year long.) And – oh, the palm trees!

We are using our intelligence to create technology, airplanes, cozy homes with everything to comfort us there, and yet we are destroying our planet. In spite of our growth in so many areas, humans are destroying the Earth..

So where I'm going with this is: I'm appealing to each and every one reading this, let's get AWAKENED to what's going on with our beautiful planet and do our absolute BEST to inspire and empower others to do the same and realize that ***we, all together should STOP the destruction*** that is happening ! Let's bring all our talents to the table, all across the globe *NOW,* ***and make the changes that are so needed.***

That is my hope, with all my heart, that ***we will make the choice to go on...!***

Please join me on this journey to AWAKEN ourselves and others to save our planet.

Yours,

Ilka

CHAPTER 1

What is going on? We have problems . . . stats, facts and resources.

"We shall require a substantially new manner of thinking if mankind is to survive".

—ALBERT EINSTEIN, GERMAN SCIENTIST (1879 – 1955)

We just had a small earthquake in Michigan during May 2015 and most people thought it was just a big truck going down the street. What? An earthquake in Michigan? Unheard of! Or how about the snow/ice storm in Alabama and Georgia during February 2014 when many were traveling toward Georgia and Florida to enjoy some sunshine and escape the harsh winter but-surprise! Ice storms stranded people on the highways and driving was disastrous! Unusual weather now is becoming more of the norm…

No one is happy to experience these kinds of disasters which are wide-spread, and the severity of them is increasing – colder winters with more ice making it more treacherous to drive, hot summers with harsher sun with more solar flares that can easily lead to more sunburns and possibly more skin cancers. Let's look at what's going on in these past few days around the world: July 2015: 120+ lightenings and darkness, rain and severe winds in the Berlin, Germany area; extreme heat in Bulgaria and some of the other Balkan countries; 17 days of non-stop rain in Ohio many flooded basements there. In the Grand Rapids paper today, I read the article on deadly flooding in Flat Gap, Kentucky with missing people, cars, and other victims ... and the list of dramatically changing climate goes on and on.

What makes this worse and more sad is the fact that WE *humans* are creating this climate change by our own actions and/ or – lack of the right action(s).

We pollute, drill, frack, crack, and many times because of financial greed, *we don't take* precautions or don't change and replace old parts or machines that brake, explode, and cause disasters such as the notorious BP oil spill , oil and chemical spills in rivers and lakes as the one in Lake Papka, Florida or the, Kalamazoo River. Because of human *ignorance and greed* , many large companies and development firms do *not have a clear understanding* of the huge impact their actions will cause to our beautiful Earth and for many, many generations to come.

Non-biodegradable chemicals and pollutants are poured into our rivers and onto the soil which has a horrible and significant effect on our human hormones and on our children and grandchildren's ability to reproduce themselves to sustain life on our planet.The poisoning of our Earth is disrupting the delicate, needed balance in our animal kingdom's life and all the plants and trees, too.

We are experiencing the effects of the so called "Global Warming/ Climate Change"; the first one to talk about it that I remember was Al Gore. If you look at Wikepidia – you will read:

> **Global warming** and **climate change** are terms for the observed century-scale rise in the average temperature of the Earth's climate system and its related effects.[2]
>
> Multiple lines of scientific evidence show that the climate system is warming.[3][4][5] Although the increase of near-surface atmospheric

temperature is the measure of global warming often reported in the popular press, most of the additional energy stored in the climate system since 1970 has gone into ocean warming. The remainder has melted ice and warmed the continents and atmosphere.[6][a] Many of the observed changes since the 1950s are unprecedented over decades to millennia.[7]

Scientific understanding of global warming is increasing. The Intergovernmental Panel on Climate Change (IPCC) reported in 2014 that scientists were more than 95% certain that global warming is mostly being caused by increasing concentrations of greenhouse gases (GHG) and other human (anthropogenic) activities.[8][9][10] Climate model projections summarized in the report indicated that during the 21st century the global surface temperature is likely to rise a further 0.3 to 1.7°C (0.5 to 3.1°F) for their lowest emissions scenario using stringent mitigation and 2.6 to 4.8°C (4.7 to 8.6°F) for their highest.[11] These findings have been recognized by the national science academies of the major industrialized nations[12][b] and are not disputed by any scientific body of national or international standing.[14]

Future climate change and associated impacts will differ from region to region around the globe.[15][16] Anticipated effects include warming global temperature, rising sea levels, changing precipitation, and expansion of deserts in the subtropics.[17] Warming is expected to be greater over land than over the oceans and greatest in the Arctic, with the continuing retreat of glaciers, permafrost and sea ice. Other likely changes include more frequent extreme weather events including heat waves, droughts, heavy rainfall with floods and heavy snowfall;[18] ocean acidification; and species extinctions due to shifting temperature regimes. Effects significant to humans include the threat to food security from decreasing crop yields and the abandonment of populated areas due to rising sea levels.[19][20]

Possible societal responses to global warming include mitigation by emissions reduction, adaptation to its effects, building systems resilient to its effects, and possible future climate engineering. Most countries are parties to the United Nations Framework Convention on Climate Change (UNFCCC),[21] whose ultimate objective is to prevent dangerous anthropogenic climate change.[22] The UNFCCC have adopted a range of policies designed to reduce greenhouse gas emissions[23][24][25][26] and to assist in adaptation to global warming.[23][26][27][28] Parties to the UNFCCC have agreed that deep cuts in emissions are required,[29] and that future global warming should be limited to below 2.0°C (3.6°F) relative to the pre-industrial level.[29][c]

On 12 November 2015, NASA scientists reported that human-made carbon dioxide (CO_2) continues to increase above levels not seen in hundreds of thousands of years: currently, about half of the carbon dioxide released from the burning of fossil fuels is not absorbed by vegetation and the oceans and remains in the atmosphere.[31][32][33][34]

Since some people are still questioning it – there is a good source: *www.climate.NASA.gov/evidence*; "Climate Change – how do we know that is happening?/Vital signs of the planet; great stats, data, graphs and pictures and – all the references, too. Included there is a statement from the IPCC:

"Scientific evidence for warming of the climate system is unequivocal."

—INTERGOVERNMENTAL PANEL ON CLIMATE CHANGE

We, *Homo sapiens* (humans) are about 200,000 years old. In the early 1800's we reached our first billion in population. In the year of 1964 we had 6.65 billion people and now, as I'm writing this (9/22/15), **we are at 7,368,854,413 people** and growing!

It is more than obvious that ***we are struggling*** to keep up with enough resources of food, water and energy for all. We are depleting AND poisoning our resources.

According to *www.worldometers.info* statistics follow with bolded areas of special concern:

Our Environment

- 3,769,987 Forest loss this year (hectares)
- 5,075,423 Land lost to soil erosion this year (ha)
- 26,352,640,661 CO_2 emissions this year (tons)
- 8,699,091 Desertification this year (hectares)
- 7,098,733 Toxic chemicals released in the environment this year

Sources and info:

- Toxic Release Inventory (TRI) Program – U.S. Environmental Protection Agency (EPA)
- United Nations Environment Program
- U.S. Environmental Protection Agency

Food

- 789,558,905 Undernourished people in the world
- 1,614,307,445 Overweight people in the world
- 538,102,482 Obese people in the world
- 19,951 People who died of hunger today
- $ 317,164,623 Money spent for obesity-related diseases in the USA today
- $ 123,329,978 Money spent on weight loss programs in the USA today

Water

- 8,845 Water consumed today (billion L)
- 1,305,435 Deaths caused by water related diseases this year
- 683,046,474 People with no access to a safe drinking water source

Energy

- 255,549,616 Energy used today (MWh), of which:
- 206,994,064 – from non-renewable sources (MWh)
- 48,555,552 – from renewable sources (MWh)
- 1,908,005,959,861 Solar energy striking Earth today (MWh)
- 54,688,293 Oil pumped today (barrels)
- 1,171,980,453,280 Oil left (barrels)
- 13,952 Days to the end of oil (~38 years)
- 1,128,176,645,804 Gas left (boe)
- 59,378 Days to the end of gas
- 4,365,869,786,319 Coal left (boe)
- 150,547 Days to the end of coal

And – in case you were curious to see how carcinogens and disease impact humans, look at the statistics for ***World's Health/Causes of death***, take a look:

- 9,410,108 Communicable disease deaths this year
- 5,509,806 Deaths of children under 5 this year
- 30,452,535 Abortions this year
- 249,199 Deaths of mothers during birth this year
- 37,510,148 HIV/AIDS infected people
- 1,218,559 Deaths caused by HIV/AIDS this year
- 5,953,333 Deaths caused by cancer this year
- 711,016 Deaths caused by malaria this year
- 9,875,062,829 Cigarettes smoked today
- 3,623,669 Deaths caused by smoking this year
- 1,812,978 Deaths caused by alcohol this year

- 777,317 Suicides this year
- $ 289,984,952,466 World spending on illegal drugs this year
- 978,505 Road traffic accident fatalities this year

Share This!

If you wondered WHY lately (at least in the last few years), you go outside for a walk around the block or hiking, but really didn't feel energized like you used to, perhaps it is important to pay special attention to the toxins released into your air...

Toxic chemicals released by industries this year: 7,098,589

Sources and methods:

Facts

- Each and *every second 310 Kg of toxic chemicals are released* into our air, land, and water by industrial facilities around the world.
- This amounts to approximately *10 million tons (over 21 billion pounds)* of toxic chemicals released into our environment by industries each year.
- Of these, over *2 million tons (over 4.5 billion pounds) per year* are *recognized carcinogens*. This amounts to about *65 Kg* each second.

Definitions

- toxic chemicals: substances that can cause severe illness, poisoning, birth defects, disease, or death when ingested, inhaled, or absorbed by living organisms.
- recognized carcinogens: as defined by the Occupational Safety & Health Administration (view OSHA Carcinogens list)
- release: any spilling, leaking, pumping, pouring, emitting, emptying, discharging, injecting, escaping, leaching, dumping or disposing into

the environment (as defined by EPCRA, U.S. Emergency Planning and Community Right-to-Know Act, section, 329–8)

- environment: includes water, air and land and the interrelationship which exists among and between water, air, and land and all living things.

Sources

For worldwide statistics, we refer primarily to the data provided by the United Nations Environment Program and correlated United Nations agencies.

For the United States, the data is provided by the U.S. Environmental Protection Agency (EPA)

The main providers of data and statistics about the environment and sustainable development for the European Union are:

- Eurostat
- EEA
- EIONET
- (for more resources – SEE back of the book, too)

Author note: *I was curious about* ***the changes in the world during the past five months*** *that I have been writing this book. See below and compare to the previous statistic and you will see how the earth is being destroyed and how the population is growing at an unbelievable rate.* ***On 2/22/16:***

World Population

- 7,403,718,608 **Current World Population**
- 20,633,458 Births this year
- 242,885 Births today
- 8,640,730 Deaths this year

- 101,714 Deaths today
- 11,992,728 Net population growth this year
- 141,171 Net population growth today

Environment

- 749,688 Forest loss this year (hectares)
- 1,009,282 Land lost to soil erosion this year (ha)
- 5,305,697,198 CO_2 emissions this year (tons)
- 1,729,873 Desertification this year (hectares)
- **1,411,631 Toxic chemicals released in the environment this year (tons)**

Food

- 777,557,735 Undernourished people in the world
- 1,621,426,447 Overweight people in the world
- 540,475,483 Obese people in the world
- 18,977 People who died of hunger **today**
- $ 306,339,330 Money spent for obesity related diseases in the USA today
- $ 117,308,549 Money spent on weight loss programs in the USA today

Water

- 1,560,955,675 Water consumed this year (million L)
- 121,387 Deaths caused by water related diseases this year
- **662,240,376 People with no access to a safe drinking water source**

Energy

- 243,072,733 Energy used today (MWh), of which:
- 196,887,844 – from non-renewable sources (MWh)
- 46,184,890 – from renewable sources (MWh)
- 1,814,850,009,870 Solar energy striking Earth today (MWh)
- 52,018,207 Oil pumped today (barrels)
- 1,159,130,555,519 Oil left (barrels)
- 13,799 Days to the end of oil (~38 years)
- 1,125,268,232,731 Gas left (boe)
- 59,225 Days to the end of gas
- 4,361,427,846,352 Coal left (boe)
- 150,394 Days to the end of coal

Author note: ***And- one month later, on 3/22/16:***

World Population

- 7,410,322,966 **Current World Population** (about 7 million MORE in 1 month)
- 31,996,265 Births this year
- 250,365 Births today
- 13,399,163 Deaths this year
- 104,846 Deaths today
- 18,597,102 Net population growth this year
- 145,519 Net population growth today

Government & Economics

- $ 7,086,350,182 Public Healthcare expenditure today
- $ 6,215,762,936 Public Education expenditure today

- $ 3,073,107,695 Public Military expenditure today
- 16,372,171 Cars produced this year
- 32,036,440 Bicycles produced this year
- 51,986,129 Computers produced this year

Society & Media

- 568,347 New book titles published this year
- 315,875,194 Newspapers circulated today
- 420,921 TV sets sold worldwide today
- 3,487,160 Cellular phones sold today
- $ 127,939,414 Money spent on videogames today
- 3,332,849,931 Internet users in the world today
- 142,020,350,484 Emails sent today
- 2,732,478 Blog posts written today
- 394,482,322 Tweets sent today
- 2,948,076,297 Google searches this year

Environment

- 1,162,539 Forest loss this year (hectares)
- 1,565,092 Land lost to soil erosion this year (ha)
- 8,246,851,786 CO_2 emissions this year (tons)
- 2,682,512 Desertification this year (hectares)
- 2,189,014 Toxic chemicals released in the environment this year (tons)

Food

- 775,284,929 Undernourished people in the world
- 1,622,774,659 Overweight people in the world
- 540,924,887 Obese people in the world

- **19,561 People who died of hunger today**
- $ 316,690,488 Money spent for obesity related diseases in the USA today
- $ 120,920,854 Money spent on weight loss programs in the USA today

Water

- 2,422,858,321 Water consumed this year (million L)
- 188,234 Deaths caused by water related diseases this year
- **658,868,220 People with no access to a safe drinking water source**

Energy

250,557,718 Energy used today (MWh), of which:

- 202,950,648 – from non-renewable sources (MWh)
- 47,607,070 – from renewable sources (MWh)
- 1,870,735,027,293 Solar energy striking Earth today (MWh)
- 53,620,014 Oil pumped today (barrels)
- 1,156,697,015,353 Oil left (barrels)
- 13,770 Days to the end of oil (~38 years)
- 1,124,717,431,458 Gas left (boe)
- 59,196 Days to the end of gas
- 4,360,586,622,591 Coal left (boe)
- 150,365 Days to the end of coal

Health

- 2,901,765 Communicable disease deaths this year
- 1,699,041 Deaths of children under 5 this year
- 9,390,553 Abortions this year

- 76,845 Deaths of mothers during birth this year
- 37,969,282 HIV/AIDS infected people
- 375,763 Deaths caused by HIV/AIDS this year
- 1,835,811 Deaths caused by cancer this year
- 219,254 Deaths caused by malaria this year
- 9,682,163,641 Cigarettes smoked this year
- 1,117,419 Deaths caused by smoking this year
- 559,062 Deaths caused by alcohol this year
- 239,699 Suicides this year
- $ 89,421,751,278 Money spent on illegal drugs this year
- 301,738 Road traffic accident fatalities this year

*So – is there such a thing called **"Let's go get some fresh air"** anymore? Wow – what a scary thought, isn't it?*

Plus – BE AWARE/keep in mind that there are toxins (and carcinogens) you also get into your body thru your skincare products and everyday personal hygiene products such as toothpaste, mouthwash, soaps, shampoos and conditioners. Make sure to look for the healthy, non-toxic ones!

Did you know that according to WHO Polluted Environment Is Responsible for One-Quarter of Global Deaths ? (TakePart.com via Yahoo! NewsMar 17 12:23 PM

"Roughly 12.6 million people died in 2012 owing to working or living in an unhealthy environment, according to report released Tuesday by the World Health Organization. **Air** and water **pollution**, along with exposure to chemicals and radiation, contribute to more than 100 life-threatening diseases and injuries."

Wow ... How sad ... and – SPEAKING OF SAD: what seems *worse then anything that I ever heard of* is ... **Monsanto.**

As a foreigner, I didn't know much about it. I heard here and there, people complaining how they (Monsanto) are responsible for killing veterans (with Agent Orange), the creation of GMO's, DDT and poisoning people with those toxic chemicals; spilling chemicals and toxins in waters and being sued over. In my "foreigner's mind," I was asking myself, "If those people are responsible for killing other people (with the Agent Orange) and killing/poisoning others through chemicals/spills – WHY aren't they in jail?" (This seems fair. If you commit a crime that kills people, you go to jail, right?)

Well, not so fast! This is definitely NOT easy to convince the justice system that Monsanto is responsible for many deaths. I remembered the movie "Erin Brokovich," and have continued to learn about the power of politics and lobbyists. When researching for this book, I was shocked when reading that Monsanto started 100+ years ago and wow! The horrible things they have done not just in the USA but worldwide! People have been marching and protesting against them all over the world.

The cold hard truth again – as some people explained to me here: "Profits are put BEFORE people," and I guess they meant "*people's* lives". Go figure that one out – I ran away from the Communism 25 years ago to come to WHAT?

If I could speak to Monsanto, I'd say:

> "WHY you do all those horrible things? Stop killing us, because you do kill your own ... children, family, grandchildren, too. Do you think they would forgive you for cutting their future short and destroying their (and **our- everybody's) home/our planet?** Monsanto – your children will die young, because of YOU, poisoning the air they breathe, contaminating the waters they might want to enjoy for boating or swimming, the soils they eat from, and – by the way – even if someone might be (forced) to clean these waters, there are still sediments settled on the BOTTOM of that lake, or go on the bottom of the contaminated river and most of the time those toxins are non-biodegradable (they'll NOT dissolve for 100+ years) and they will be passed from the fish that eat them to the next creature(s)

that eats THOSE fish (including us humans) and on and on. The toxins end up in our bodies, messing up our hormones and – MOST IMPORTANTLY your children's (and , of course – all others') **ability to reproduce themselves (having healthy children)!***

*Explained in more details by Dr. John Lee, MD's books (an expert on hormone imbalance).

You could make so much $$$ in a positive, inspiring way by partnering with innovators. Jump in on all the safe, good things that many have been doing and – there is so much more needed to be done! Your children and many others will be so proud of you! And, yes – people are forgiving and willing to give someone a "second chance" **for the greater good of all!** Think about it and – **do it ...** It will be great ... You don't need to even say, "I'm sorry!" – I know, it might be a difficult thing to do, but you can just "take a turn" , do that change inside and you'll be rewarded in such a great way! In so-ooooo many ways!) And – you'd inspire OTHER people and companies to follow you!

Well – I hope I'm not too late with my appeal, because I just heard the NEWS and found THIS in late December 2015:

Monsanto Faces People's Trial for Crimes against Humanity

Posted on Dec 3 2015 – 6:25pm by Sustainable Pulse

Greenpeace/France human health/International Criminal Court/ Monsanto/Tribunal Foundation/Products/Soil/World Food Day

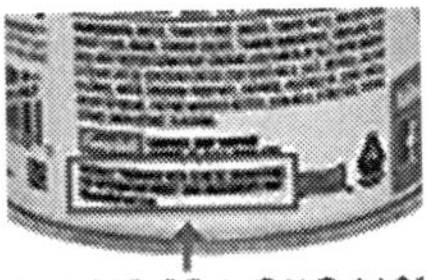

XAMPLE OF A GMO LABI

EU Commissioner Admits New Multi-Herbicide GM Soybeans need Health Safety Assessment

Campbell Soup Set to Disclose GMOs in Shock to US Food Industry

Brazil Justice Ministry Fines Nestle over Failing to Label GMOs

Monsanto Cuts 16% of Work Force as Sales in Roundup Herbicide Fall 34%

GMOs Set to Crush Export Markets for Farmers across Ghana

Source: www.monsanto-tribunal.org

The Organic Consumers Association (OCA), IFOAM International Organics, Navdanya, Regeneration International (RI), and Millions Against Monsanto, joined by dozens of global food, farming and environmental justice groups announced today **that they will take Monsanto MON (NYSE), a US-based transnational corporation, to a people's trial for crimes against nature and humanity, and ecocide, in The Hague, Netherlands, next year on World Food Day, October 16, 2016.**

Since the beginning of the twentieth century according to the groups, Monsanto has developed a steady stream of highly toxic products which have permanently damaged the environment and caused illness or death for thousands of people. These products include:

- PCBs (polychlorinated biphenyl), one of the 12 Persistent Organic Pollutants (POP) that affect human and animal fertility;
- 2,4,5 T (2,4,5-trichlorophenoxyacetic acid), a dioxin-containing component of the defoliant, Agent Orange, which was used by the US Army during the Vietnam War and continues to cause birth defects and cancer;
- Lasso, an herbicide that is now banned in Europe;
- and RoundUp, the most widely used herbicide in the world, and the source of the greatest health and environmental scandal in modern history. This toxic herbicide, designated a probable human carcinogen by the World Health Organization, is used in combination with genetically modified (GM) RoundUp Ready seeds in large-scale monocultures, primarily to produce soybeans, maize and rapeseed for animal feed and biofuels.

Relying on the "Guiding Principles on Business and Human Rights" adopted by the UN in 2011, an international court of lawyers and judges will assess the potential criminal liability of Monsanto for damages inflicted on human health and the environment. The court will also rely on the Rome Statute that created the International Criminal Court in The Hague in 2002, and it will consider whether to reform international criminal law to include crimes against the environment, or ecocide, as a prosecutable criminal offense. The International Criminal Court, established in 2002 in The Hague, has determined that prosecuting ecocide as a criminal offense is the only way to guarantee the rights of humans to a healthy environment and the right of nature to be protected.

The announcement was made at a press conference held in conjunction with the COP21 United Nations Conference on Climate Change, November 30 – December 11, in Paris.

Speaking at the press conference, Ronnie Cummins, international director of the OCA (US) and Via Organica (Mexico), and member of the RI Steering Committee, said: "The time is long overdue for a global citizens' tribunal to put Monsanto on trial for crimes against humanity and the environment. We are in Paris this month to address the most serious threat that humans have ever faced in our 100–200,000 year evolution – global warming and climate disruption. Why is there so much carbon dioxide, methane and nitrous oxide in the atmosphere and not enough carbon organic matter in the soil? Corporate agribusiness, industrial forestry, the garbage and sewage industry and agricultural biotechnology have literally killed the climate-stabilizing, carbon-sink capacity of the Earth's living soil."

Andre Leu, president of IFOAM and a member of the RI Steering Committee, said: "Monsanto is able to ignore the human and environmental damage caused by its products, and maintain its devastating activities through a strategy of systemic concealment: by lobbying regulatory agencies and governments, by resorting to lying and corruption, by financing fraudulent scientific studies, by pressuring independent scientists, and by manipulating the press and media. Monsanto's history reads like a text-book case of impunity, benefiting transnational corporations and their executives, whose activities contribute to climate and biosphere crises and threaten the safety of the planet."

Marie-Monique Robin, journalist and author of the best-selling documentary (and book by the same name), "The World According to Monsanto," said: "This International Citizens' Tribunal is necessary because the defense of the safety of the planet and the conditions of life on Earth is everyone's concern. Only through a collective resurgence of all living forces will we stop the engine of destruction. That's why today I am calling on all citizens of the world to participate in this exemplary tribunal."

Also speaking at the conference were Valerie Cabanes, lawyer and spokesperson for End Ecocide on Earth; Hans Rudolf Herren, president and CEO of the Millennium Institute, president and founder of Biovision, and member of the RI Steering Committee; Arnaud Apoteker, creator of the anti-GMO campaign in France, which became one of the priority campaigns of Greenpeace France, and author of "Fish in Our Strawberries: Our Manipulated Food;" and Olivier De Schutter, co-chair of the International Panel of Experts on Sustainable Food Systems (IPESFood) and former UN Special Rapporteur on the Right to Food.

More information is available at www.monsanto-tribunal.org/

Fracking ... another greed-based problem facing our earth!

Some people thought, "Oh, there is nothing bad about it – it has been done for so many years." Really? And that means it is not damaging the earth? Well, let's take a look:

"High volume horizontal hydraulic fracturing – commonly called fracking – was developed in the late 1990s.[1] After drilling the well several miles underground, first vertically and then horizontally, companies mix chemicals – many of them cancer-causing or neurotoxic – with millions of gallons of water and sand, then blast the mixture underground under intense pressure to break up deep shale or other rock formations and extract oil and natural gas." ★

(Source: www.letsbanfracking.org); check it out for more info on the harms of fracking.

And – who is polluting THE MOST in our world?

If you want to see MORE on the *top10 polluters, here it is:*

6 Graphs Explain the World's Top 10 Emitters; *www.wri.org/profile/mengpinge by* Mengpin Ge, Johannes Friedrich *and* Thomas Damassa – *November 25, 2014*

China currently ranks as world's largest emitter of greenhouse gases.
Photo by Ding Zhou/Flickr

Our ability to harness ever-expanding amounts of data is completely transforming our understanding of environmental problems and solutions. Our Climate Insights blog series *leverages data from* CAIT 2.0, *WRI's climate data explorer, to shed light on the many dimensions of climate change that shape society, policy, and global development.*

Recently published the CAIT Equity Explorer can help assess the equitability of these pledges. While equity considerations include more than just emissions – and all countries must take action to mitigate climate change – the actions of the world's top emitters will likely be most heavily scrutinized – and rightly so: According to recent data, 10 countries produce around 70 percent of global GHG emissions.

and the people denying we have a problem with Climate Change…

First I thought these were the ones (in Congress) that deny Climate Change because they were paid to do so by the biggest polluters: oil, gas and other energy extractors that destroy our lands close to home, state and national parks by polluting the waters, the air, and other lands by chemical spills, leaks, dumps and fracking. So, the *"polluters"* support the political campaigns of the *"deniers"*; sort of, "You scratch my back, I'll scratch yours"! This is what I've been hearing and reading…

"Deniers", really? Is this the best you got? What is the price you are paying for the $$$ you are getting? You say that humans are not creating Climate Change and that corporations are NOT to blame but you are ruining the lives and the future of your own children and grandchildren. By contributing to the destruction of YOUR world, the wind will blow toxic chemicals in your yard and that of your children and grandchildren.

Right? Think of all the cancers because of the toxic pollutions (or – melanomas, leukemia) and the ozone layer thinning providing us with highly intense sunshine leading to more skin cancer.

If we destroy our Mother Earth, do you think we can move or fly to another planet? Well, so far **"We don't have a planet "B"** (As one senator mentioned in a testimony with scientists in Washington, DC.) So stop denying the truth and siding with BIG BUSINESS and the constant polluters. I have also read about scientists who are also "deniers" and have gone to the 'dark side' – they have been paid big $$$ to deny corporate polluting leads to Climate Change.

CHAPTER 2

How our poisoned Earth is impacting us back

"Challenges are the Universe's way of saying: Get creative!"

—MARY BAILLEE

How is global warming effecting everything? Right now we have total imbalance on Earth. Please read on.

Our Water

People depending on fishing do not want poisoned fish for dinner! It's more and more rare to declare, "We have clean waters (oceans, seas and rivers) to eat from and to enjoy such as swimming." There are very few pristine areas of water in the world. So many bodies of water are being poisoned by river runoff from corporations.

I love swimming. This summer I swam in Duck Lake near Lake Michigan and got the worst ear infection! I had never had an ear infection before and now I'm upset that I can't swim in small lakes anymore as they are polluted! I appeal to you – ***please voice your opinion*** and *fight for your rights* to have CLEAN water – no more oil (or other toxic) spills, and massive toxic runoffs from corporations into our rivers and streams.

Of course, the BP oil spill was the most notorious and caused unbelievable destruction and harm to so many animals that live in or near the water including humans, birds, fish, and wildlife that depend upon the water for their food and sustenance. The damage will be felt for decades to come!

The nonprofit World Wildlife Fund (WWF) and the Zoological Society of London have jointly determined that *pollution, climate change and industrial-scale over-fishing* **killed half of all marine life over the last 40 years**. Large swaths of coral reefs, mangroves and sea grasses have also died, further decimating fish populations. Statistics show how the family of fish that includes tuna and mackerel has declined by 75% since 1970. The number of other species is also declining; 25% of all shark and ray species face extinction. I know many of us might not like shark in particular, but all water creatures including shark are needed for balance.

French biologist Gilles Boeuf says that ***½ of all coral*** *has already disappeared*, and the rest will vanish by 2050 if temperatures continue to rise at current rates. *(Source:Tinyurl.com/WWF2015BluePlanetReport)* Isn't that horrible? Coral are (in my opinion) *the most beautiful part of the oceans or seas! What are we doing? We are destroying all the beautiful things in the water and on Earth.*

People living on *small islands* ARE BECOMING COVERED with water as we speak. In the National Geographic's Climate issue, Nov. 2015 the article *"Against the Tide" (how to live with it)*" by Kennedy Warne , talks about the small islands of Kiribati and the people living on them – the I-Kiribati. They lived in a paradise of *marava, karawa and tarawa* – the sea, land and sky for many, many years, but now "the trinity is tilting out of balance."

The sea (or "Mother Ocean") is NOT that "heart of providence the people have always known." They are faced with climate change and the uncertainty and fear that comes with it. The world keeps telling them that "low-lying island countries like theirs will be soon under water."

Their own leaders have said that Kiribati – 33 coral islands (with a capital Tarawa atoll) in an expanse of the central Pacific larger than India – is "among the most vulnerable of the vulnerable".

But even so, many I-Kiribati refuse to think of their homeland as a "disappearing island nation" and of themselves as "sinking islanders", but rather as "descendants of voyagers, inheritors of a proud tradition of endurance and survival".They don't want to think that their paradise is lost, yet to many of them it seems so unfair that they struggle with Climate Change troubles and didn't contribute to it even a bit!

As the writer Kennedy Warne mentions, *"Since the 1980's, Pacific leaders have scolded, cajoled, pleaded with , and tried to shame the major carbon-polluting countries over climate change. The islands are ants and industrialized nations are elephants",* declared Teburoro Tito, a former Kiribati president. Talks of drowning islands and *climate refugees* has made Kiribati known around the world.

Other atolls at risk from the rising seas are Maldives, Marshall Islands TokelauandTuvalu.A formerTuvaluan prime minister , Saufatu Sopoaga, went so far as to compare the impacts of climate change to *"a slow and insidious form of terrorism against us."* Can you blame these leaders sharing their concern? You know the answer. Even if these islanders didn't contribute to the climate disaster, they are WILLING to do something about it and not be a "defeated people". They love their homeland (or – island I should say),and are willing to do anything they can to stay, rather than relocate. I, as an immigrant myself, can say that is not easy to leave the country you lived at least half of your life in. I admire those people for their spirit of resilience and not wanting to be "defeated people".

A young man from the Paris Summit this month and – sharing this with the Sierra Club – got up and talked to the people there about how their village in Alaska has been eroded for 35 years and now is even worse: people are forced to leave their villages and homes because of the effects of the Global warming and in particular – the ice melting. I'm quoting the e-mail I got from the Sierra Club (with their permission):

Dear Ilka,

I'm writing you from Paris where the United Nations climate talks are underway. It's a long way from my hometown of Shishmaref, Alaska, population: 650. But there's a good reason I felt compelled to be here.

Shishmaref is a barrier island that has been eroding and flooding for the past 50 years, even before climate disruption was widely recognized. Over the past 35 years, we've lost between 2,500 and 3,000 feet of land. Within the next two decades, our whole island is likely to erode away completely and become uninhabitable.

While in Paris, I am determined to speak up for my community of Iñupiaq and for all native peoples disproportionately affected by climate disruption around the world. Please join me in calling for a strong and equitable climate agreement.

To my community, strong climate action includes taking ambitious steps to mitigate carbon pollution, enable vulnerable communities to relocate safely, and contribute to the Green Climate Fund for climate adaptation and loss and damage.

In 2001, my people voted to relocate along the coast of mainland Alaska. It really hurts knowing that my only home is going to be gone, and we won't hunt, fish, and carry on traditions the way that we're accustomed to and that our people have done for centuries.

Even though we made this decision, I don't think anyone is really prepared. The older generations want to stay on the island because they've lived here their whole lives, and they don't want to leave their homes. Everyone wants to stay, but we are beginning to realize that we have no choice except to relocate and we need a strong climate agreement and government help to do that.

As a young person, it's important to me that a climate agreement, not only corrects for decades of persistent pollution, but also prepares us for the future. One thing my community knows all too well is that we can't afford to wait any longer. Paris is our best opportunity to act.

With hope from Paris,

Esau Sinnok

Sierra Student Coalition

Indigenous Environmental Network

Arctic Youth Ambassador

P.S. Want to follow the Sierra Student Coalition's youth delegation in Paris? Follow @sierrastudent on Twitter.

This email was sent to: **russiantalk@gmail.com**
This email was sent by the Sierra Club
85 2nd St San Francisco, CA 94105

Paris Climate Talks: What's at Stake for the World's Coasts

Infographic (on next page) outlines how sea level rise triggered by climate change will permanently alter coastlines and how much the resulting damage will cost.

By Zahra Hirji and Paul Horn

Dec 4, 2015

People travel on a boat as they move to safer places through a flooded road in Chennai, India on December 2, 2015. India is among several countries that will have the highest number of people affected by flooding as global warming accelerates. REUTERS/Stringer

Editor's Note: *This is the first infographic in the ICN "What's at Stake" series highlighting the key impacts associated with climate change.*

Climate change is already redrawing the world's coastlines – and it will continue to do so as long as the earth continues to warm. According to a recent analysis by Climate Central, continuing to burn fossil fuels at the current rate would result in warming of 4 degrees Celsius compared to preindustrial levels. That would cause nearly 30 feet of sea level rise, submerging the land that's home to at least 470 million people if no adaptation measures are taken. Here's a rundown of how climate change affects the coasts and the associated social and economic costs...

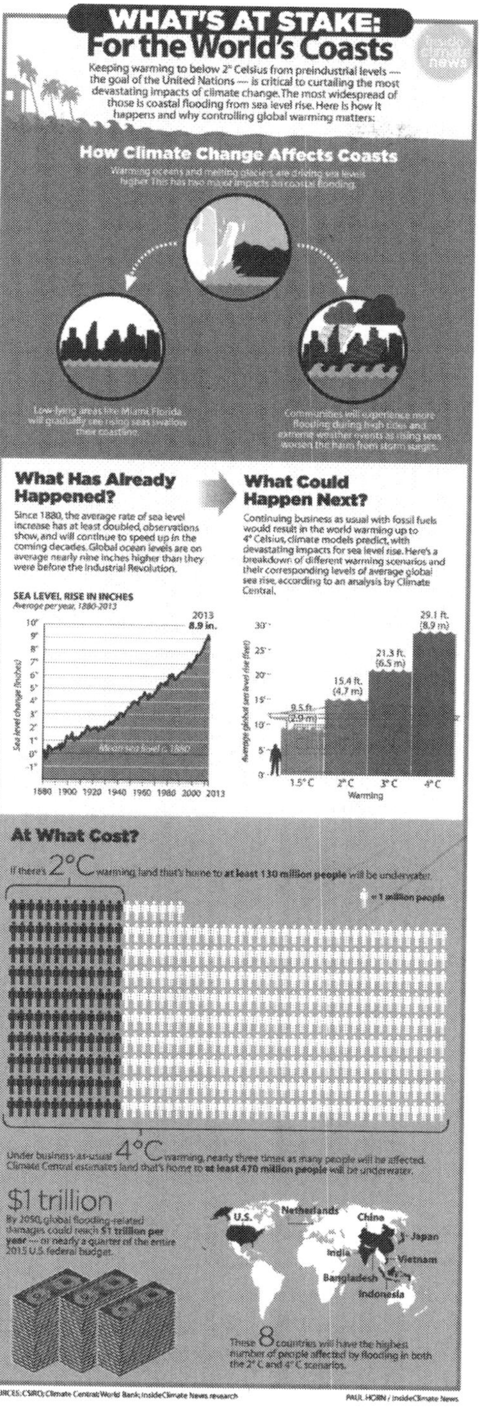

SOURCES: CSIRO; Climate Central; World Bank; InsideClimate News research

PAUL HORN / InsideClimate News

InsideClimate News is a Pulitzer Prize-winning environmental news organization whose investigative reporting exposed that Exxon knew decades ago that fossil fuels cause climate change.

* NASA's research/dangers for NASA's centers

There is the real danger for all the NASA facilities, since they all are close to water.*:

Sea Level Rise Hits Home at NASA

Watching Waters Rise Right Outside the Front Door

The Sun rises at Kennedy Space Center. The sea is steadily doing the same. (Photo by NASA/Andres Adorno) By Michael Carlowicz Design by Joshua Stevens & Paul Przyborski August 26, 2015

(They are supposed to be close to water because in case of failure of rockets being launched or other aircraft experiments – better to be above water, so they don't endanger people and our cities).

And as the author of this article explains, "Sea level rise hits especially close to home because half to two-thirds of NASA's infrastructure and assets stand within 16 feet (5 meters) of sea level. With at least $32 billion in laboratories, launch pads, airfields, testing facilities, data centers, and other infrastructure spread out across 330 square miles (850 square kilometers) – plus 60,000 employees – NASA has an awful lot of people and property in harm's way."

"As the ocean has warmed, polar ice has melted, and porous landmasses have subsided, global mean sea level has risen by 8 inches (20 centimeters) since 1870. The rate of sea level rise is faster now than at any time in the past 2,000 years, and that rate has doubled in the past two decades."

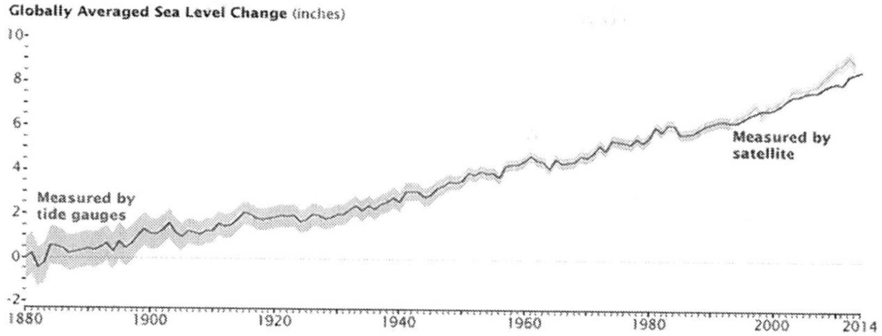

Tide gauges have been used to measure sea level for more than 130 years. Satellite measurements now complement the historical record. (NASA Earth Observatory image by Joshua Stevens, based on data from the Commonwealth Scientific and Industrial Research Organization and NOAA)

The maps here show (in red) the areas around NASA centers that would be inundated by 12 inches (30 centimeters) of sea level rise. The table below shows the recent trends and predicted sea level rise at several NASA centers. According to global climate models (GCMs), sea levels are expected to rise at least 5 inches (13 centimeters) for most NASA centers by the 2050s. Those numbers rise significantly if ice sheets continue to melt rapidly.

In a review of the agency's vulnerability to sea level rise, NASA's Climate Adaptation Science Investigators (CASI) Working Group recently wrote:

Sea level rise of between 13 and 69 centimeters by the 2050s is projected for NASA's five coastal centers and facilities...

Air Pollution

- Factories pumping clouds of CO_2 (carbon dioxide) in the air you breathe; this is causing more breathing difficulty amongst humans and creating pollution-caused respiratory illnesses.

- Pollution from the cars we drive. Please buy eco-cars and drive when absolutely necessary. Consider walking or biking, when possible.
- China is the #1 polluter globally. The GOOD News is that China seems to be doing something positive towards reducing the carbon emissions.

> "The world's biggest producer of carbon emissions is seeking to boost the use of renewable fuels as smog has blanketed cities from Shanghai to Beijing, forcing factories and schools to close and intensifying pressure on public officials to cut pollution. This month China suspended price adjustments for fuel as a way to curb automobile exhaust and it has pledged to peak carbon emissions around 2030, by which time it aims to derive 20 percent of the energy it uses from clean sources."
>
> "Beijing officials on Friday raised the city's air pollution alert to orange, the second-highest on the four-grade scale, warning children and the elderly to avoid outdoor activities as limited visibility from the thick smog forced the airport to cancel 227 departures. The chronic air pollution has renewed calls for the government to make better forecasts and act faster to help clear the skies over the city of 21.5 million."

"Beijing Raises Smog Alert as Airport Cancels 227 Departures "

—BLOOMBERG NEWS DECEMBER 24, 2015 – 8:04 PM EST
UPDATED ON DECEMBER 25, 2015 – 3:54 AM EST)

Red Alerts

You might have heard of ***China's Dec. 7th Red Alert***. And then ***the 1/6/2016 Red Alert!*** If you Google *"China's red alert* for air pollution" – you will see a lots of news coverages, including *www.Foxnews.com/video*: "China issues first ever "red alert" for air pollution, where Benjamin Hall, a reporter from London covers it, saying that the RED level means "severely harmful for all" and this fact really shocked me: **1.4 million people die prematurely annually from smog in China...** Wow – can you imagine that? Really sad ... and – yes – "they put the growth of their economy first" ... but at least there is some movement to the right direction.

Weather disasters as a result of destroying Mother-Earth's balance:

- ***tsunamis,*** because of earthquakes
- ***earthquakes*** because of drilling anywhere we can, fracking and whatever other experiments we can possibly come up with – ALL ***affect*** Earth's balance, directly and indirectly. Scientists and experts say "Climate change affects some weather, but we shouldn't blame it for every extreme event". (Source: "Climate Change Survival Guide: How to Live with it" in the National Geographic's Nov. 2015 Climate issue. *That issue, by the way, is one of the BEST articles put together on Climate Change with scientific studies about how this affects us, and how it is affecting the animal kingdom, including the marine life. Great work, National Geographic! Thank you for that!)*
- ***tornadoes*** – we need to adjust to these realities and prepare/create shelters
- ***hurricanes*** because of warming oceans
- ***floods*** because of rising of the sea levels (hurricane Sandy, 2012 in NY)
- ***droughts*** and chronic water shortages harm rural areas and city dwellers.
- ***Health risks/disasters***: result(s) of high heat, rain and higher humidity will produce more infectious diseases : more ticks, so more Lyme disease, more allergies, more respiratory disease; mosquito viruses, including a NEW KIND-Zika virus. Bill Reisin, an entomologist of the University of California Davis confirms that "the thermodynamics of mosquitoes are driven by temperature".

From an article in Grand Rapids Press, "Higher temperatures help mosquitoes transmit Zika", by Seth Borenstein, 2/4/16. Zika was found in a pregnant woman that brought it from Brazil. That new mosquito-borne virus, plus the well-known ones: West Nile virus, and the Asian flu virus which can produce effects including malaria, diarrhea and malnutrition. The list goes on and on … and will affect us ALL including our children and grandchildren if we allow the global warming to continue!

CHAPTER 3

The Good: good ideas, solutions, people and organizations, local and world leaders working and supporting the efforts/making a difference.

"Never doubt that a small group of thoughtfully committed citizens can change the world. Indeed, it's the only thing that ever has".

—MARGARET MEAD

It is about time for some Good News, right? And, someone actually gave it a name (abbreviation): *TIISG*- "Turn It Into Something Good", Dr. Joe Vitale.

Good news: The hole in the ozone layer is finally starting to heal

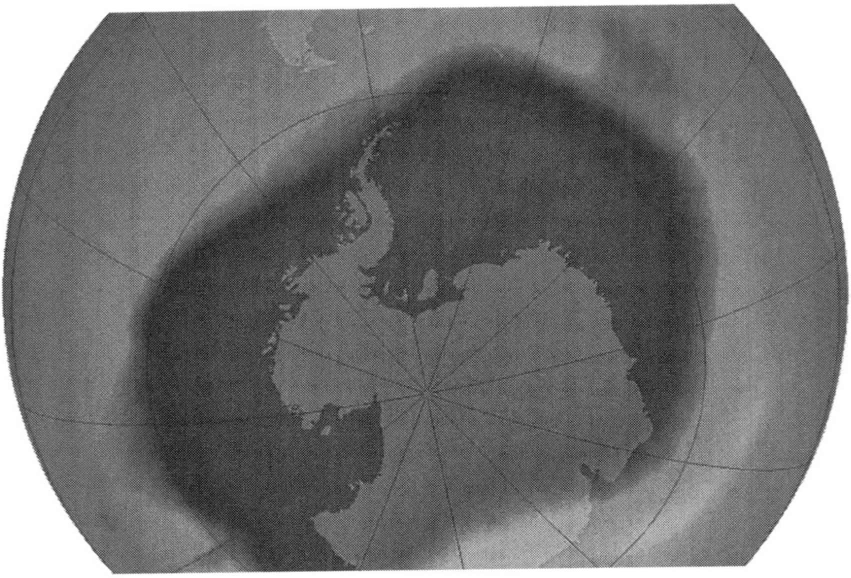

Image of the largest Antarctic ozone hole ever recorded (September 2006), over the Southern pole NASA

Sometimes the world really *can* get together and avert a major environmental catastrophe before it's too late. A new UN report finds that the Earth's protective ozone layer is finally starting to recover – after efforts in the 1980s to phase out CFCs and other destructive chemicals.

Scientists have finally detected signs that the ozone layer is healing

China, USA and India (known as the biggest polluters in the world) are doing something to improve the pollution/CO_2

China – "China will stop approving new coal mines for the next three years and continue to trim production capacity as the world's biggest energy consumer tries to shift away from the fuel as it grapples with pollution.

China will suspend the approval of new mines starting in 2016 and will cut coal's share of its energy consumption to 62.6 percent next year, from 64.4 percent now, Xinhua News Agency reported Tuesday, citing National Energy Administration head Nur Bekri. ***It's the first time the government has suspended the approval of new coal mines, according to Deng Shun, an analyst with ICIS China.***

The world's biggest producer of carbon emissions is seeking to boost the use of renewable fuels as smog has blanketed cities from Shanghai to Beijing, forcing factories and schools to close and intensifying pressure on public officials to cut pollution. This month China suspended price adjustments for fuel as a way to curb automobile exhaust and it has pledged to peak carbon emissions around 2030, by which time it aims to derive 20 percent of the energy it uses from clean sources.

"This new policy, along with efforts to eliminate inefficient mines, may help to ease the severe domestic oversupply" of coal, Deng said by phone from Guangzhou. "It will take several years to take effect."

Renewable Boost

The country will also close more than 1,000 coal mines next year, taking out 60 million metric tons of unneeded capacity, according to the Xinhua report. China shuttered a similar number of mines this year, wiping out 70 million tons of production, according to a separate statement from the NEA dated Dec. 29. The country is on track to produce 3.58 billion tons of coal this year, down 0.5 percent from 2014, according to the NEA.

China plans to increase wind and solar power capacity by more than 21 percent and have at least 20 gigawatts of new wind power installations and 15 gigawatts of additional photovoltaic capacity next year, according to the NEA statement."

Source :Bloomberg News, December 29. 2015

India and Paris Agreement What COP21 holds for India

The Paris Agreement "*is hailed by many developed and developing countries including US and India as a landmark & historical moment.*"

Surprisingly, there is more skepticism than optimism among developing nations that the developed nations will need to transform their coal-dependent economies to one that is more sustainable and have reliance on cleaner energy resources.

India, has been commended by many developed nations including the United States & France for its role in the climate summit. India came a long way ever since the whole debacle in the last climate summit, agreement has more growth prospect for India this time without any serious harm.

It is to be noted that India has always been committed to engage actively in multilateral negotiations in a positive and forward-looking manner. Indian negotiators have emphasized on establishing an effective and equitable understanding based on climate justice and the principle of Equity and Common, but Differentiated Responsibilities and Respective Capabilities, under the United Nations Framework Convention on Climate Change (UNFCCC).

One of the major achievements for India was that it was not forced to declare a peaking year for its Carbon emission as China did last year during its Joint talks with United States.

This means that India will have no restrictions on its energy consumptions and thus it will not affect its Industrial development; which was feared by many environmentalists before the commencement of summit. Secondly there is no sanctions on usage of Coal by developing nations, which is a major relief for India considering that 78% of its electricity generation is through coal and a sanction or limited usage warning due have led to major energy crisis in the country.

BERTRAND GUAY/AFP/Getty Images

The Article 4 of the agreement, clearly states that "Parties aim to reach the global peaking of greenhouse gas emissions as soon as possible, recognizing that peaking will take longer for developing countries. Thus the signed agreement clearly acknowledges that developing countries including India should get longer time to clean up its emissions as any other developed countries.

The COP21 agreement provides India with vast opportunities to develop industries domestically in areas like solar energy, electric vehicles, etc. Though India lacks the technological knowhow, but has the advantages of being an underdeveloped market which has huge growth potential and incentives for industry leaders to set up operations in here. The saturation of US, EU and China for energy and automobile sector also works well in India's favor.

While working towards a sustainable environment, one should remember the great words of Mahatma Gandhi **"One must care about the world one will not see"**. Indeed, humanity has progressed when it has collectively risen to its obligation to the world and responsibility to the future.

The Paris Summit/COP21 (the 21-st Conference of the Parties

For me, just the fact that the world got together to DO SOMETHING about the problem we all have – is a step forward/promising and gives us all hope!

So speaking of that, here in Grand Rapids, Michigan, we had our own **ex-Mayor George Heartwell,** that was at the Paris Summit and spoke to the public and shared with all of us who were interested in hearing what happened in Paris. He spoke at a few different venues, and brought some excitement, hope, and the fact that ***progress was made!***

There were 37 other Mayors from the USA; 600 from all around the world.The event was described mostly as *"monumental"*, and *"the process was promising". Each country was required to present their goals PRIOR to the event.!*

And here is the **MOST IMPORTANT section of the Paris Summit:** the elements of the agreement, graciously shared with me by our ex-Mayor, George Heartwell:

Here are the bones of the Agreement:

- First, every participating nation – 196 of us – was required to submit its goals for Green House Gas reduction in advance of the conference. This was an enormous breakthrough. Neither the important Kyoto Climate Change Agreement of 1997 nor any of the subsequent update agreements had required nations to set measurable, verifiable goals of their own. I'll say more later about how that happened. What is important to note for now is that the Agreement has an appendix setting forth the goals established by every nation.

- Second the agreement sets a long-term goal of maintaining global temperature average increase at no more than 2 degrees Celsius (which is 3.6 degrees Fahrenheit) above temperatures of the pre-industrial era. Furthermore, the Agreement states that nations will "pursue efforts" to keep the rise in temperature to no more than 1.5 degrees Celsius. By the year 2050, the Agreement holds that man-made emissions should be reduced to a level that the forests and oceans can naturally absorb.
- Third, beginning in 2020 and every five years thereafter nations will be required to report their progress toward goals and establish new goals. You see, the cumulative effect of all the nations' current goals reported in the Paris Accord still result in a temperature rise above 2 degrees. So, it is imperative that we get progressively more aggressive in goal setting.
- Fourth, there is also a requirement that the developed nations provide financial support to the less-wealthy developing nations to allow them to "leapfrog" energy technologies and make other investments in emissions reduction that might not otherwise be possible. To illustrate, let's say that a developing nation needs to build a power plant to keep up with its development pace and fossil-fuel electric production technology is the least expensive to construct and operate. That nation can secure grants from a fund to be established by the developed nations to "leapfrog" the old technology and invest in cleaner, though more expensive, technology. The amount of funding required was not written into the agreement but the stated goal is $100B/yr. This will require a significant increase from our nation and others. Addressing the Mayors of the world in Paris City Hall, France's President Holland pledged to increase his nation's foreign aid contribution from its present $2.0B/yr. to $5.0B. America needs to be similarly bold and proportionally greater.

With respect to China and India the important pre-COP21 commitments that each made to emissions reductions, China by Central Government mandate and India by Cap and Trade system – along with President Obama's Clean Power Plan – set the stage for other nations to be more aggressive in setting their own goals."

George Heartwell was a great Mayor and a leader of the city of Grand Rapids, indeed! Just the fact that he served three terms (2004-2016) – shows how much people loved him and appreciated his leadership! Actually – he could have been a Mayor again, if there wasn't "a term limit vote" – now the limit is 2 terms/8 years.

In case you want to know how great he was as Mayor (*he is STILL continuing* to work on the environmental theme) – here are some comments on his Grand Rapids *achievements during his terms:*

> "Most of you know the Grand Rapids story. You know that a dozen years ago a Mayor – called a "tree-hugger" by even his friends, worse by his opponents – set out to make us the most environmentally, socially and economically sustainable city in the nation. You probably know that the United Nations designated Grand Rapids its *first Regional Center for Expertise in Sustainable Development in the US*, that the *US Chamber of Commerce recognized Grand Rapids as the nation's Most Sustainable Mid-Sized City*, that the *US Conference of Mayors gave Grand Rapids the first place climate protection award in the large city category* because our efforts surpassed those of cities substantially larger than we. Perhaps you know *that the EPA ranks us in the top 20 cities for renewable energy* and that *the US Green Building Council recognizes that we have more LEED certified buildings per capita than any other city in the nation*. And you know that we did this in one of the more politically conservative regions in the country."

At his talks, George Heartwell mentioned, that "it starts with each one of us, to raise awareness locally. The primary needs start at a local level". Our current Mayor Rosalynn Bliss is continuing this great legacy as she is driving sustainable practices and she is committed to renewable energy.

⋆Our president, Obama – his ***Clean Power Plan*** is well-liked by many; he, most likely, gained and keeps gaining more supporters, by doing his best to address the Climate Change issue. Here is his ***LAST State of the Union speech on Jan. 12, 2016:***

Obama on Climate and Energy in 2016 State of the Union By ICN Staff Jan 13, 2016

President Obama speaks to Secretary of State John Kerry after ***Obama completed his final State of the Union address on January 12, 2016.*** Credit: REUTERS/Jonathan Ernst

Here are the parts of President Obama's 2016 State of the Union address that were about ***energy and climate change,*** taken from the text prepared for delivery:

> Look, if anybody still wants to dispute the science around climate change, have at it. You'll be pretty lonely, because you'll be debating our military, most of America's business leaders, the majority of the American people, almost the entire scientific community, and 200 nations around the world who agree it's a problem and intend to solve it.
>
> But even if the planet wasn't at stake; even if 2014 wasn't the warmest year on record – until 2015 turned out even hotter – why would we want to pass up the chance for American businesses to produce and sell the energy of the future?

> Seven years ago, we made the single biggest investment in clean energy in our history. Here are the results. In fields from Iowa to Texas, wind power is now cheaper than dirtier, conventional power. On rooftops from Arizona to New York, solar is saving Americans tens of millions of dollars a year on their energy bills, and employs more Americans than coal – in jobs that pay better than average. We're taking steps to give homeowners the freedom to generate and store their own energy – something environmentalists and Tea Partiers have teamed up to support. Meanwhile, we've cut our imports of foreign oil by nearly sixty percent, and cut carbon pollution more than any other country on Earth.
>
> None of this will happen overnight, and yes, there are plenty of entrenched interests who want to protect the status quo. But the jobs we'll create, the money we'll save, and the planet we'll preserve – that's the kind of future our kids and grandkids deserve.

It's great to have a President and a leader that supports what matters *most* to you, isn't it?

Climate Change and President Obama's Action Plan

The United States is leading global efforts to address the threat of climate change. On December 12, 2015, President Obama and world leaders secured a global agreement to combat climate change. **Learn more now.**

> ***The "Mission Innovation"*** **,** "an initiative to dramatically accelerate public and private global clean energy innovation to address global climate change, provide affordable clean energy to consumers, including in the developing world, and create additional commercial opportunities in clean energy."

(Source: November 29, 2015 at 7:01 PM ET by Paul Bodnar, Dave Turk)

Paul Bodnar is the National Security Council's Senior Director for Energy and Climate Change, National Security Council. Dave Turk is Deputy Assistant

Secretary for International Climate and Technology at the U.S. Department of Energy.

Summary: President Obama joins world leaders in announcing "Mission Innovation" to dramatically accelerate global clean energy innovation to address climate change.

"Today in Paris, President Obama and French President Hollande, along with a wide range of other top global leaders, will announce "Mission Innovation."

Through the initiative, 20 countries are committing to double their respective clean energy research and development (R&D) investment over five years. These countries include the top five most populous nations – China, India, the United States, Indonesia, and Brazil. They stretch across five continents. And when you add all partner countries together, they represent 75 percent of the world's CO_2 emissions from electricity, and more than 80 percent of the world's clean energy R&D investment."

And once you start reading – you'll find also more on the "parallel initiative", called ***"Breakthrough Energy Coalition",*** spearheaded by Bill Gates – more of a Private Sector – led effort that has pledged to invest extraordinary levels of private capital in clean energy, focusing on early-stage innovations".

I see people working all together for our MOST-WORTHY cause ever: to sustain our Home (the Earth)!

Did you know that in October 2015, Secretary of State John Kerry brought investors together at a Climate and Clean Energy Investment Forum to develop better pathways to sustainable development in new and emerging markets, including finding new ways to accelerate collaboration across borders and sectors? ***Earlier this summer, Vice President Biden* announced $4 billion in independent commitments *by major foundations, institutional investors, and others to fund innovative solutions to help fight climate change,*** including technologies with breakthrough potential to reduce carbon pollution.

See more details on all of the above at:
www.whitehouse.gov/the-press-office/2015/06/16/fact-sheet-obama-administration-announces-more-4-billion-private sector

– plus – you'll see (at the end) THIS:

Photos of beautiful, clean Mother Nature at her finest.

Clean Power Plan

Americans have demonstrated their dedication to climate action through a wide variety of commitments. Learn more about these efforts.

The Clean Power Plan

The Clean Power Plan sets achievable standards to reduce carbon dioxide emissions by 32 percent from 2005 levels by 2030. By setting these goals and enabling states to create tailored plans to meet them, the Plan will:

Protect the health of American families. In 2030, it will:

- Prevent up to 3,600 premature deaths
- Prevent 1,700 non-fatal heart attacks
- Prevent 90,000 asthma attacks in children
- Prevent 300,000 missed workdays and schooldays

Boost our economy by:

- Leading to 30 percent more renewable energy generation
- in 2030
- Creating tens of thousands of jobs
- Continuing to lower the costs of renewable energy

Save the average American family:

- Nearly $85 a year on their energy bills in 2030
- Save enough energy to power 30 million homes in 2030
- Save consumers $155 billion from 2020–2030

Pope Francis

Now: this is EXCITING! A world-figure who is changing the world wherever he goes, whenever he speaks. I call him "our loving Father", because *this is who he is.* You hear and see this man and you can't help but to feel the love, and the compassion, and the inspiration exuding from him and the concern he has for our "Common Home" as he calls the Earth. There was so much excitement in the USA when he was here last September (9/15). People traveled from all over the country to come and see him, hear him speak; maybe – feel the words of hope and faith that "the good" always prevails. There were people around me talking while we were discussing what he is doing and saying how much they fell in love with him, even if they were not Catholics. Because it is all about the love, the common desires we share, no matter what religion we believe in.

> So with Pope Francis it was all natural for a "loving Father" like him to absolutely get involved in responding to our biggest problem (issue) we have right now ... **Climate Change.** Before his visit to the USA in Sept., 2015 *Pope Francis had a meeting at the Vatican with about 60 mayors from around the world who were invited "because they back Pope Francis' environmental message" and gathered on July 23 for 2 days* "to keep the pressure on world leaders ahead of the Paris negotiations in December. "The mayors signed a declaration later Tuesday (7/21) that stated ***"human-induced climate change is a scientific reality and its effective control is a moral imperative for humanity".***
>
> *(Source: Nicole Winfield, The Associated Press; her article "Mayors seek climate agreement" in the Grand Rapids press of 7/23/15)*

As many of you know, Pope Francis wrote his *encyclical* ***"On Care for Our Common Home" (Laudato Si)*** where he denounces the "fossil-fuel-based world economy that exploits the poor and destroys the Earth". He invites us all to dwell upon the truth that it's all connected: the environment is relevant to our faith and as Michael Wright is saying in a *"Catholic Update"* from September 2015 "Its care is tied to respect for God's creation. Along with issues of life , family, and the poor, care for creation is a principle of Catholic social teaching."

I don't know of a better comparison of *WHAT we are doing to our environment*, but comparing **it** to ***sin***. And that is what Pope Francis does when quoting Patriarch Bartholomew of Constantinople:"***For human beings to degrade the integrity of the earth by causing changes in its climate, by stripping the earth of its natural forests or destroying its wetlands; for human beings to contaminate the earth's waters, its land, its air, and its life – these are sins".*[15] *For "to commit a crime against the natural world is a sin against ourselves and a sin against God".*[16] *(8) (from our Pope's encyclical).***

Well, this is exactly what I wrote in my diary on 8/4/15 (about the Pope):"*I'm so excited that the Pope is so serious on working on the problem of our Climate change! Way to go! Yes, our "loving Father", our religious leader is absolutely recognizing the fact that we have a big problem and we need to address it ASAP, NOW!"* (then I'm mentioning the Mayors' meeting with the Pope).

And the more of us that see that world-figures (as the Pope) are involved – *the MORE of us will get inspired to do something about it.* We are the ones that made this"mess", so we should be the ones to take care of it. But especially when we are inspired,, we start to believe MORE that even *one person can make a difference*! And that is SIGNIFICANTLY important!

Leonardo DiCaprio

Another world-figure: the amazing actor drew huge attention indeed when he spoke about climate change in his Oscar-winning speech. People were roaring with excitement inside the Hollywood theatre, and also all over the globe with tweets and pins, "Congrats!" and "Great job!" "Wow! Thank you, thank you, thank you, Leo for addressing the most urgent issue we have"!

What a man! I hope many get inspired TO ACTION by him; yes, that's right, we NEED ACTION; talk is cheap. When world-figures like that speak up for what they believe and inspire us to action, we should ACT. If duplicated by millions (or even billions) of people, imagine BIG difference!

I quote this one from The Christian Science Monitor:

> Leonardo DiCaprio uses Best Actor Oscar speech to draw attention to climate change (+video) DiCaprio won the 2016 Oscar for Best Actor for his role in the film 'The Revenant.' 'Climate change is real,' the actor said during his acceptance speech. 'It is happening right now.'
>
> By **Molly Driscoll, Staff writer** February 29, 2016

"The Revenant" actor Leonardo DiCaprio won an Oscar for Best Actor after having been nominated several times before and DiCaprio ***used a large amount of his time onstage to draw attention to climate change.***

He won his Academy Award for playing Hugh Glass, a real-life figure who was injured by a bear in the wilderness and had to survive in difficult conditions.

OSCARS 2016: Leo acceptance speech highlights climate change

> " 'The Revenant' was the product of the tireless efforts of an unbelievable cast and crew I got to work alongside," the actor said. " ... Mr. Alejandro Iñárritu, as the history of cinema unfolds, you have forged your way into history these past 2 years."

DiCaprio then brought up the topic of climate change, mentioning the trouble those behind "Revenant" had in finding an area with snow to be able to film the movie.

> "Climate change is real," the actor said. "It is happening right now. It is the most urgent threat facing our entire species, and we need to work collectively together and stop procrastinating. We need to support leaders around the world who do not speak for the big polluters, but who speak for all of humanity, for the indigenous people of the world, for the billions and billions of underprivileged people out there who would be most affected by this. For our children's children, and for those people out there whose voices have been drowned out by the politics of greed. I thank you all for this amazing award tonight. Let us not take this planet for granted."

For those who are familiar with the actor, you know that he has been involved with the cause of "making the world a better place" for many years now. And for those wanting to know: he co-founded the Leonardo DiCaprio Foundation (leonardodicaprio.org) in 1998 and produced the 2007 climate documentary "The 11th Hour." Last month, the actor discussed the topic with Pope Francis. He also appeared before the UN climate summit in 2014 to ask for a move forward on the topic.

In *www.leonardodicaprio.org* you can find so much more info on all his inspiring-the-world-to-action work, including how in January, Leonardo DiCaprio attended the World Economic Forum in Davos Switzerland to receive the prestigious Crystal Award. The award is given to artists and cultural leaders who are helping to address the world's humanitarian and environmental challenges.

Here is an exciting report by Divest Invest, and their movement, which DiCaprio supports:

> NEW REPORT CHARTS STUNNING GROWTH OF DIVESTMENT MOVEMENT & ANSWERS FAQs
>
> "With over 500 organizations worth more than $3.4 trillion pledging, the growth of the Divest-Invest movement in 2015 is without precedent. A new report by Divest-Invest Philanthropy and Europeans for Divest Invest examines the history and success of this dynamic movement, and provides answers to frequently asked questions about the implications for the institutional investor."
>
> Divest-Invest Philanthropy welcomed its first members in January 2014. Since then, more than 125 foundations are walking the talk, while beating their benchmarks. Hans Joachim Schellnhuber, Papal adviser and director of the Potsdam Institute for Climate Impact Research calls divestment "the most important action that ever happened on climate change."– LDF
>
> As the actor said: "This is the most urgent of times, and the most urgent of messages."

> And he walks his talk by "flying on commercial planes rather than private jets, and reportedly has solar panels on his house. He bought a Toyota Prius in 2006. Last year, DiCaprio was plugging the [electric sports car the] Tesla Roadster."
>
> *(Quote from CS Monitor, By Molly Driscoll, Staff writer February 29, 2016)*

Al Gore

Let's not forget the man who started all that movement and concern!

Many years ago, when the reality of the Global Warming just started and we heard about Al Gore, there were two harsh winters and I remember many people saying ,"Yep, so much for global warming! It's brutally cold!!!" But we have been learning more and more since…!

Al Gore has a lifelong commitment to reversing the effects of global climate change. He has been a longtime advocate for the environment. Gore was elected to the U.S. House of Representatives 4 times (since 1976), and the U.S. Senate in 1984 and 1990. He was inaugurated as Vice President of the United States on January 20, 1993, and served eight years. Gore spends the majority of his time as chairman of The Climate Reality Project, a non-profit devoted to solving the climate crisis. He is the author of the bestsellers *Earth in the Balance, An Inconvenient Truth, The Assault on Reason, Our Choice: A Plan to Solve the Climate Crisis*, and most recently *The Future: Six Drivers of Global Change.* "He is the subject of the Oscar-winning 2006 documentary film "An Inconvenient Truth", directed by Davis Guggenheim about former United States Vice President Al Gore's campaign to educate citizens about global warming via a comprehensive slide show that, by his own estimate made in the film, he has given more than a thousand times." (From Wikipedia)

The former Vice president is also the co-recipient, with the Intergovernmental Panel on Climate Change, of the 2007 Nobel Peace Prize for "informing the world of the dangers posed by climate change."

I think Al Gore became even more famous on the subject after his movie "An Inconvenient Truth" with his message to all: "It is now clear that we face a deepening global climate crisis that requires us to act boldly, quickly, and wisely." I sincerely HOPE we really do so.

For more info – go to: *www.algore.com*; Wikipedia; Huffingtonpost.com, etc.

Bill Clinton

He was the 42nd U.S. President from 1993–2001

After just one month in office, on February 22 1993, President Clinton outlined his priority: "Long-term economic growth that creates jobs and *protects the environment*."

A significant contribution of the Clinton Administration to environmentalism was that trade agreements between the United States and other nations required that those countries observe environmental standards.

After his presidency, he founded The **Clinton Foundation** (founded in 2001) as the **William J. Clinton Foundation**, and then called (during 2013–15) " the **Bill, Hillary & Chelsea Clinton Foundation**[5])" (from Wikipedia): It is a non-profit corporation with the stated mission to ***"strengthen the capacity of people throughout the world to meet the challenges of global interdependence."*** The Foundation focuses on improving global health and wellness, increasing opportunity for women and girls, reducing childhood obesity and preventable diseases, creating economic opportunity and growth, and helping communities address the effects of climate change. The Foundation works principally through partnerships with like-minded individuals, organizations, corporations, and governments, often serving as an incubator for new policies and programs. Its offices are located in New York City and Little Rock, Arkansas.

The Clinton Foundation consists of few different entities, including the Clinton Health Access Initiative (CHAI, spun off into a separate but related organization in 2010), the Clinton Global Initiative (CGI, split off after 2009), Clinton Global Initiative University (CGI U), the Clinton Climate Initiative (CCI), and few more, including the No Ceilings Project.

I was impressed to find out that "Through 2014, the Foundation had raised almost $2 billion from U.S. corporations especially Wall Street; foreign governments and corporations; political donors; and various other moneyed interests.[3] During its lifetime the foundation has received praise from philanthropic experts and has had support from both Democrats and Republicans". (Wikipedia)

Through the Clinton Global Initiative 2,300 Malawian farmers in the Trees of Hope program have planted more than 2, 6 million trees, sequestering more than 200,000 tons of CO_2!

And *they invite us to help* "Fight poverty and climate change with Trees of Hope"!

$25 helps to offset 2.5 tons of CO_2 emissions. A lots of great programs, projects, etc., including the fact that more than 31,000 American schools were provided with healthy food choices in an effort to eradicate childhood obesity and – over 450,000 people have been impacted through market opportunities created by social enterprises in Latin America, the Caribbean, and Asia; through the independent Clinton Health Access Initiative, 9.9 million people in more than 70 countries have access to CHAI-negotiated prices for HIV/AIDS medications and so much more … (for more – go to clintonfoundation.org/clintonglobalinitiative.org)

As a keynote speaker at Enviro-Expo in Windsor on 4/16/2010, Bill Clinton said: "The Earth is alive. It's a tangible, living thing. We, as part of that breathing thing, need to take care of it". One thing I like about Bill Clinton is his energy of "cooperation." I was reading an article in our Grand Rapids Press: *"Bill Clinton, the anti-fighter pol"* by Tim Skubick and that was addressed again as he was speaking to over 600 FOB (Friends of Bill) crowd at MSU: "Mr. Clinton was all about bipartisan cooperation as he retold the story of Nelson Mandela, the South African civil rights advocate who went from jail cell to leader," the author said. And talking about USA he said,"The only thing that bothers me about American politics, is everybody thinks nobody will vote for them if they are not fighting with somebody ... What works is ***creative cooperation.* Constant conflict is a loser."**

Organizations, making a difference ("Movers and shakers"):

Sierra Club

A great organization for protecting our natural resources. Their logo is simple: ***"Explore, Enjoy, and Protect the planet",*** so in order for us to enjoy all the beauty of our planet – we are simply asked to be voices for Earth!

Here is an example of an e-mail I got from the Sierra Club (after I joined them). It's so simple: join them and you'll be getting these emails with news and be a VOICE to protect and safe our Earth! Take a moment to show the BLM you support protecting our climate from methane pollution, and urge the Obama administration to end all fossil fuel leasing on public lands.

> Dear Ilka,
>
> The oil and gas industry is profiting from digging up and fracking our public lands.
>
> To add insult to injury, the industry is dumping huge amounts of climate-harming methane emissions into our atmosphere, either by letting it pour out freely or by burning it (commonly referred to as "flaring") at drilling sites on our public lands.
>
> That's why **the Bureau of Land Management (BLM) has proposed important new standards to crack down on this harmful gas flaring and venting**.
>
> **Take a moment to urge the administration to quickly finalize these rules to protect our climate and our public lands.**

One important part of this solution is to establish strong standards to limit harmful emissions coming from oil and gas infrastructure already operating on our public lands today. This is what the BLM is seeking to do with this proposed set of rules to rein in methane pollution from gas venting and flaring on our public lands.

Take a moment to show the BLM you support protecting our climate from methane pollution, and urge the Obama administration to end all fossil fuel leasing on public lands.

The BLM's proposed rules are a big step in the right direction; however regulating methane is not enough to make fracking safe for our climate or communities. Ultimately, the most effective way to solve the climate crisis and to protect our communities is to keep all fossil fuels in the ground. Together we'll keep moving towards that goal.

Thanks for protecting our climate and communities,

Lena Moffitt

Director, Sierra Club Beyond Dirty Fuels Campaign

This is how you can help our Earth, by joining organizations like Sierra Club that HAVE *been already doing good* for many years (for 124 years – in this case☺); you can voice your desire to save our planet, sign petitions for President Obama to STOP dirty fuels FAST! Do it for you, your children and their children! More then ever our world needs your action NOW!!!

The National Geographic Society

Is a global non-profit membership organization which says: *"We inspire thru exploration, illuminate thru stories, and, always, teach"*. You probably enjoy their magazines (National Geographic); their logo says it all: "We believe in the power of science, exploration and story telling to change the world".

I love their magazines as they are so full of beauty – the Earth everywhere … on every continent! And of course: all the creatures that inhabit it, from the animals, to the birds, and all the marine life. Great informational articles, too. Their November 2015 issue, ***"The Climate Issue"*** is one to absolutely explore: with so much information, facts and data, innovation and ideas, and great articles!

CREDO Mobile (CredoAction)

Another organization supporting great causes, asking for your help to be a VOICE and support these causes for your own good; here is one example of their e-mails:

Tell President Obama: Time to get tough on environmental carcinogens

> Petition to President Obama
>
> "The federal government is failing to adequately protect us from toxics, chemicals and pollutants in our air, water, food, homes and workplaces, which are a largely unaddressed contributor to cancer. If you want to do everything in your power to cure cancer, it's time to take steps to follow the 2010 recommendations of the President's Cancer Panel, to get tough on environmental carcinogens."
>
> Add your name:

Another good email was how to protect the bees from the pesticides that are killing them. We all know how important the bees are for us to have all the foods that come from the crops and the fruits and veggies (besides the honey☺)

Be a voice! YOU and millions of others like you and me can make things happen the way we want them! Individuals to organizations, every-day people like you and me ARE making changes more then you realize!!!

CHAPTER 4

Innovations, innovators and ideas to save the planet

"If you think you are too small to have an impact, try going to bed with a mosquito in the room."

—ANITA RODDICK, ENVIRONMENTAL ACTIVIST

Innovations and innovators.

Have you heard of **"The Ocean Cleanup"?** I saw an article about it in our local Grand. Rapids Press from Aug. 25, 2015*: "Boaters mapping Pacific garbage arrive in San Francisco port,"* by Olga Rodriguez. I saw more pictures and info on *www.inhabitat.com*: about how a **19-Year-Old Develops Ocean Cleanup Array That Could Remove 7,250,000 Tons Of Plastic From the World's Oceans.** The article follows:

19-year-old Boyan Slat (now – 21 years old) has unveiled plans to create an Ocean Cleanup Array that could remove 7,250,000 tons of plastic waste from the world's oceans. The device consists of an anchored network of floating booms and processing platforms that could be dispatched to garbage patches around the world. Instead of moving through the ocean, the array would span the radius of a garbage patch, acting as a giant funnel. The angle of the booms would force plastic in the direction of the platforms, where it would be separated from plankton, filtered and stored for recycling.

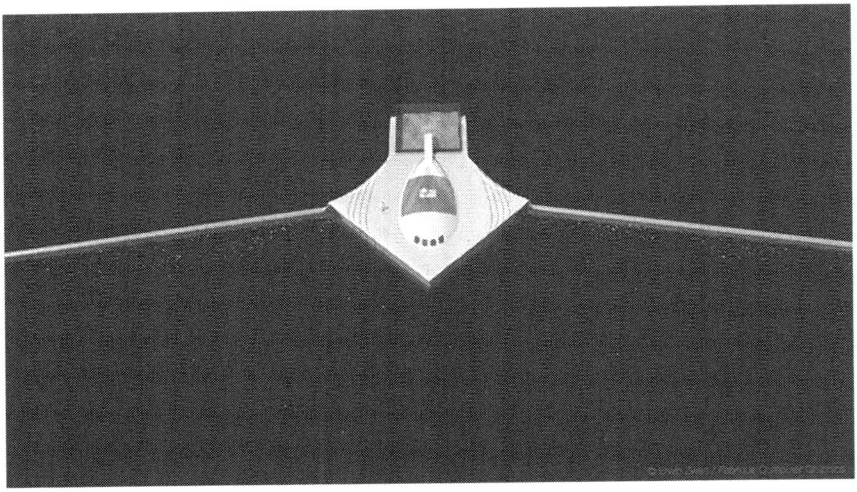

Slat went on to found The Ocean Cleanup Foundation, a non-profit organization which is responsible for the development of his proposed technologies. His ingenious solution could potentially save hundreds of thousands of aquatic animals annually, and reduce pollutants (including PCB and DDT) from building up in the food chain. It could also save millions per year, both in clean-up costs, lost tourism and damage to marine vessels.

It is estimated that the clean-up process would take about five years, and it could greatly increase awareness about the world's plastic garbage patches.

Caltech/Resnick Institute is looking for ... *you!*

In the Article ***by Bob Silberg, "Caltech calls for "Crazy New Science to save the planet",*** he states how

> "Delia Milliron wants your windows to stop being so stupid. Mika Järvinen wants to lay waste to the concept of waste. Joel Dawson wants the world's cell phone systems to save the energy equivalent of 7 million cars, while Yi Cui wants to put you into a shiny new Tesla."

Got your own crazy idea to save the world? Caltech's Resnick Institute just might honor you next year with its Resonate Award, a tribute it bestowed on each of these scientists this year. (*www.resnick.caltech.edu/awards.php*)

> "Of course, there's crazy and then there's crazy. The kind the Resnick Institute is looking for is born of great intellect uninhibited by convention. "Truly disruptive innovation can turn on its head our preconceived notion of how the world can be," said the institute's director, Harry Atwater, as he introduced the 2015 award winners at a panel discussion held in Aspen, Colorado, this past July."

The Great Energy Challenge

National Geographic and Shell partnered:

Great innovations, great ideas and info; and visionaries whose bright ideas have been powered by grants through them … See it all at *www.greatenergychallenge.com*

Many countries and states use the power of wind NOT just on the mainland, but also miles away IN THE sea/ocean or in lakes where the winds are the strongest.

(For example – Germany, off the North Sea and Baltic Sea.)

And another great thing that Germany did was committing to NO NUKES: after the disaster in Fukushima, Japan, in 2011, Germany vowed to quickly abandon *nuclear energy. Of 17 operable reactors, 9 have since been shuttered..* The rest are set to close by 2022.

Europe is … ahead of the game about clean energies

I was traveling through Amsterdam the Summer of 2015 and, when landing, the first thing I saw were dozens and dozens of wind turbines, nicely lined up on a perfectly green field! It was a beautiful "Welcome to Europe."

There are so many more ideas depending on your lifestyle, preferences and *how far* you want to go to save the planet.. Here are just few, general ideas (if you wish); share and send me yours!☺

Ideas that YOU can implement in your daily life:

- hang clothes to dry: indoors or outside
- wash your laundry in cold water – if you do – your household can save 1,600 lbs of CO_2 emissions/year! (Source: National Geogr. Mag., the Climate Issue, 11/15)
- stop water while soaping your body/shampooing your hair – no need for it to run

- turn off water when you brush your teeth; you are wasting water – remember the 658, 868,220 (as of today, 3/22/16) people in the world who HAVE NO access to a safe drinking water source.
- in the winter, when is 20 F outside – do you need to shower/wash you hair EVERY DAY? Maybe not...
- by all means – drive your car ONLY if needed: pile all your errands/ shopping in 1 day (if possible); great especially for self-employed and retired people. "Leaving your car at home 2 days a week can reduce your greenhouse gas emissions an average of 2 tons a year!" (Source: National Geogr. Mag., The Climate Issue, 11/15, p. 24)

Can you imagine the difference we can make if ½ of our world's population could do this? ½ of 7 billion people by 2 tons = 7 billion tons less pollution!

Wow – you see what it means that it starts with each of us, and that 1 person can make a difference.

- bathe babies OUTSIDE in your backyard in the summer (if you can:): use rain barrels (or other containers to collect rain water) – they LOVE it!
- teach your children to recycle and – not to waste food, as well as water and other precious resources – teach children and teens to go out and play ball or hockey or whatever they like – it's healthier for them, it's energizing and keeps them more focused to study and – . it's FREE! (vs. the video-games, pc, etc.) And one of the problems w/ TV/video-games, pc, phones and microwaves is that they are called "electro-magnetic pollution", since they fracture your energy field (see MORE on that in the "Appendix").
- buy an electric car if you can ... – great clean idea!
- as many of us seen on TV – the Tiny Houses trend is getting more and more popular – they're cheaper and less is MORE (more savings, more freedom: you can take them anywhere with you!) – MORE FUN. They give you a chance to use the $$$ you saved to travel and explore the world.

This is the right time **to share the idea of *"Less is MORE"*.**

I always wonder, "Why people don't see the big picture?" All they worry about is their own self-fullfiling needs..." When they will awaken to see that they are hurting the Earth?

"Oh, I got to work for a company that pollutes the Earth, but ... they are going to pay me BIG $$$! So I can have bigger a home (so more – more lumber, more trees; or – bigger car = **more pollution.)**

Our own selfish needs are causing the destruction of the Earth.

LESS IS MORE: more clean forests, more clean air, more simple lives."

Here – we seem to think alike with someone :

> *"There is a great need for the introduction of new values in our society, where bigger is not necessarily better, where slower can be faster, and where less can be more."*
>
> —GAYLORD NELSON

Got it?

I really NEED to address the **WASTE NOT issue when it comes to food and water – our 2 main staples of life.**

Food

I shared the stats (in Ch. 1) about how many people die from hunger every day (almost 20,000). And that nearly 800 million worldwide suffer from hunger.And – ⅓ of the planet's food goes to waste ... Shocking? Yes, indeed.In the article **"Waste Not, want Not" Elizabeth Royte (Nat. Geogr. Mag., 3/2016 issue)** she talks about it and states the fact that "according to the Food and Agriculture Organization of UN, we squander enough food – globally, 2.9 trillion lbs a year – to feed every one of person more than twice over." And continuing to explain WHERE is all that food going , where it is lost. (lack of proper storage facilities, refrigeration, and bad roads while transporting produce – in developing countries and – in developed nations such as the USA.

We waste down the supply chain when retailers put more than needed on our plate at restaurants or when we forget to eat leftovers at home or – simply, when we bought more than we should and toss it when it goes bad. It's shocking how much waste we have here in the USA when someone like me, coming from a different (not so developed country) notices this. We were raised to hardly ever throw food away! Every summer, I remember how we were putting fruits in glass jars with water and a bit of sugar (closed tightly) after boil dozens of fruits and veggies on a fire in our back yards (I think it's called "canning" here; we call it "compots"). Every family was doing it so we could eat in the winter.

There are 6 billion lbs of U.S. fruits and veggies unharvested or not sold because they were "ugly" or I call it not-so-perfect to be sold! Come on now! Why can't we buy and eat those? We are NOT perfect either, but no one is "throwing us away", right? They have the same nutritional value as the other ones ... Just have some fun with them, joke (with your kids),"Oh , I'm eating a 'twin' carrot or 'a kidney' tomato, etc.

The former president of Trader Joe's Doug Rauch opened a nonprofit supermarket in Dorchester, Massachusetts (DailyTable) where he sells discounted fruits and veggies and inexpensive *surplus* goods.

Water

This is our other precious source/staple of life.

As of today, March 22, 2016, six hundred million people in the world HAVE NO access to a safe drinking water source. And, actually there are "wars" for water ... (read article, for example in our Grand Rapids Press: ***"Over the line?" by Garret Ellison*** *from 1/17/16; "Wisconsin city puts Great Lakes Compact to the test).* I'm sure you probably heard about the Flint water crisis and as I was finding more about it, I heard *Erin Brokovich saying in an interview that "We should not politicize water: water should be good for everyone: Hispanic, white or black." So, let's not waste it…*

CHAPTER 5

Let's analyze this: Human dysfunction = the Earth's dysfunction or Human's change = Climate change.

"The person who says: "It cannot be done" shouldn't interrupt the one doing IT"

We reflect our imperfection onto Mother-Earth by...

A) abusing it in every way: polluting it, suffocating it – the air, the water, the soil; deforestation, fracking, etc., I call that "*the direct effect*". The other one is the *"unconscious effect"*. All because human greed is the opposite of *respect: doing business in a way so we don't harm others and nature,* **How we are: our thoughts are polluted, so polluting the Earth is just a reflection of who we are.**

B) confusing *it: because the climate change creates a weather pattern change;* it is a ripple effect, *confusion in* Mother-Nature causes crazy weather patterns. And

C) The not-as-well-known yet phenomena: we reflect our emotions and behaviors *onto Mother-Nature:* our anger, fears, the desire for destruction or craving power, acting irresponsibly, etc., also known as negative emotions or "issues in the tissues".

In other words, the way we treat ourselves or others, is the same way we treat Mother Earth with all of its creatures: the plants, the trees and animals. I think many of us *don't even realize* this, do we? It's unconscious.

This subject is HUGE … and aVERY IMPORTANT one because we are talking about things that ARE NOT taught in school YET (that I know of). But many of us have been learning and discovering (actually rediscovering, since that science has been available for a while) the *SUBCONSCIOUS*.

The best way that someone ever described it is:" *Until the SUBCONSCIOUS becomes CONSCIOUS it will rule your life and you'll call it "faith"* – Dr. Joe Vitale (Author of "The attractor factor", "Zero limits", etc.) Acting of the *opposite* of what we really intend!

You probably have heard this before, "We are all ONE!" But, did you think about what it means? I believe it means is that *what 1 person does- can affect ALL (of us)*. And that is exactly what is happening with Climate Change now.

For example – you know how much the BP's oil spill affected … all. Or how the methane leak in CA (Porter ranch) was still going 100+ days after. That impact is HUGE! ***Again – "human dysfunction" causing "Earth's dysfunction" and destruction.*** And it is because 1 person (most likely) that didn't do their math or whatever needed to have a plan B; having a way how to plug this pipe in case of this disaster. Was it greed again? Not wanting to invest in safety measures most likely. Is GREED destructive? Yes, you bet. So many lives are affected in this case and – it's still going (as I'm writing this) unfortunately.

And it might sound strange if we think about it, but it starts with these thoughts and questions: "Are we AWAKENED *to understand all of this and that* we are ALL ONE…!

Are *you* awakened? Or on the path of awakening? Do you know already why you are *here?* What is your purpose, mission, or in other words, what you are put on this Earth for? If you are already fulfilling your life's purpose – GREAT! Help others to awaken to do this too, because we are all put here to be ***joyful and happy*** and to remember and remind each other WHY we are here.

So when we go thru hardships, it's just to get stronger and be worthy of and appreciative of the beautiful life we have and the planet we share. To remember that everything good we have in life doesn't come easy, but it's so worth the struggle to gain.

And if you feel a bit grumpy sometimes – it's all natural: it's part of your *"growth period"* (I learned this from Mindvalley training – Jeffrey Allen)* or rebirth; you are rebirthing yourself every time you learn more and more about yourself! And, always remember WHY you are suffering and – that you are NOT the ONLY one. This last one gives us some peace, right? Knowing that *we are not the only ones.*

"Your own Self-realization is the greatest service you can render the world".

—RAMANA MAHARSHI

I don't know of anyone else explaining this better, but one of the best teachers I ever had/have (I'm a life-time student☺), Barbara De Angelis, in her book "Soul Shifts," said how we were passed a great wisdom from teachers who wrote about this a thousand years ago: *"There comes a profound moment in the life of a soul when the individual suddenly begins to have what they describe as the "thought of awakening": "I should learn more about the truth of life. I think I am here for something, to do something, to search for something. I must find out. I need a teacher. I need wisdom. I must wake up."*

Thank you, Barbara, my beloved teacher, for your wisdom!

Speaking of *"searching"*, if you have read *Dr. Frankl's book "Man's Search for Meaning"* – you get the idea. *Read it! It's one of the most important books you could ever read!* He explains there how research found that ***"man's search for MEANING is the primary motivation for his life"*** **...** NOT $$$, or career.

I believe that ***if*** anyone who puts POWER or $$$ first, explores what we are talking about here – they might *"shift"* or might *"move"* inside of their soul.

As one genius said it: ***"Nothing happens until something MOVES."*** (Have I *MOVED* you yet?) Let's go back to WHY! Why do we perpetrate the crazy/horrible thing regardless of our intentions to do otherwise? "WHY"? – I always believe that WHY is such a powerful word, because when we ask ourselves this question we can discover a world of truths; **stop** and analyze and think more rationally in the moment and be exactly what we humans are designed to be: intelligent, "thinkers", and then "doers". Emotions without rationality in the moment can get us in trouble. Dr. Goldman describes it that way: ***"Success in life is more dependent on our ability to manage our emotions then on our IQ.".***

Wow! How true this is! We must get in touch with ourselves and our conscious to make the right decisions in life and ask ourselves "Why am I doing this?" That would be an interesting reflection for the humans who are destroying the earth!

For each of us, we must look back at our own childhood and our heritage. Do we have family in the past who were champions for others who were challenged and who fell upon hard times? Did your parents reach out to help others heal and survive? Or was a parent selfish with a 'I'll do as I wish" mentality and you watched this as a child and began to copy that same behavior?

I wish that we could get into the minds and background of those humans who deforest our beautiful trees just so their companies can have profits for their shareholders or the CEO can have another million dollar raise. I wish we could understand what heritage or DNA is within a human who manufactures and distributes toxic products that will seep into the Earth's soil. Or the person who knows that there are toxic chemicals in a skincare product but hopes to have a large commission check when they gain national distribution of the product.

Hopefully, you the reader, have within your DNA a gene that gives you the confidence and courage to stand up to those perpetrators who are destroying our planet. Write a letter, make a call, start a campaign against the evils that continue to ruin our lives and make a living from the destruction of the planet.

Ask yourself about your own personal decisions. Question what you purchase for yourself and your family. Take a stance to support companies that are dedicated to the good of humanity and the Earth.

Learn to be good to your own mind. Harbor positive thoughts and release the toxins within. The better you are to yourself, the better you will be toward others. The body is energy. Pure and simple. We are ALL energy. One of the greatest gifts we can give ourselves and our world, it to work through negative emotions and thoughts. Consider where they came from and work toward a peaceful solution.

Focus on the good for yourself, your fellow humans, animals, nature and our land. Emit positive vibrations and energy. Give this gift to yourself and others. BE AWAKENED to who you are and what you stand for!

CHAPTER 6

Being responsible/taking responsibility for our own actions, emotions and behaviors.

"When you blame another, you give you own power away..."

—LOUISE HAY

As mentioned in the previous chapter – it is for our own good to be responsible people: we can't blame others for what is our own outcome in life, no matter what happened in our past. We cannot change the past, but we can learn from it and take steps to heal the future.

There is no better way to describe this, but one of my teachers (internationally known), Louise Hay; just nailed it:

> "When you blame another, you give your own power away because you're placing the responsibility for your feelings on someone else. People in your life may behave in ways that trigger uncomfortable responses in you. However, they didn't get into your mind and create the buttons that have been pushed. Taking responsibility for your own feelings and reactions is mastering your "ability to respond." In other words, you learn to consciously choose rather than simply react".
>
> From her book "I Can Do It".

When we don't take responsibility – we don't have any power. We feel like, "Oh, I didn't create this ..." and choose not to do anything about the problem. So, do you want to feel empowered or weak? For me that means: not willing to take action, so being ignorant is not doing the right thing and being indifferent, which is not acting upon something that you should!

Life is all about choices and it is so important to train ourselves to make the right ones. There are many people who disregard other human beings – they text recklessly when others are passengers in the car, or drink and drive. I call this ***"total disregard for human life"***...including your own!

People that respect themselves would tend to want to respect Mother Nature as well.

But many have NO respect for it. I went for a walk this past winter and the snow was slowly melting. Underneath the snow I saw lots of plastic bags, styrofoam boxes, a round can of Pringles, and a cigarette box. This was surely a sign of humans disrespecting the earth and others.

Of course, we as parents have the responsibility to teach our kids the same thing; to have respect for self, others, and the environment and this starts AT HOME.

CHAPTER 7

Being responsible for our planet as for everything else in life starts with each of us.

"Be the change that you wish to see in the world."

—MAHATMA GANDHI

You heard all of this many times: if you want things to be different – act, be a voice, be proactive! *If we do nothing – nothing (not a thing) is going to change!*

As I mentioned earlier about organizations and people that make a difference and want YOU to join them – do so! Whoever clicks with you, expresses your beliefs and values – it takes a TEAM to make a DREAM come true, right? And there is no worthier dream than to save our "home," Earth!

Let's be proactive! Write to your local Congressman, Governor or representatives, etc. for passing laws to protect the Earth with all of its natural resources, including all the National, State and any parks, actually – these are your places to enjoy, relax, get energized and happy! (And get inspired by all the beauty and peacefulness that is there for you!) Vote for the right people (locally and – for a president) who **will** do the right things!

Do not take for granted your ability to vote: it is a blessing; some people DON'T have that great right and opportunity!

I take my responsibility to vote for the right person very seriously, because that ONE person can impact the lives of our whole country (a president), but also the future of our whole world. All country leaders throughout the world should work together for the "highest good for all" of us.

A good president should be able to be a good diplomat, understand foreign policy and the World, be a "people's person", be able to listen and understand the needs of both parties (countries). Patience and the ability to manage emotions are very much needed, as this will help gain the respect of other global leaders who will listen more closely to the thoughts and actions of the president.

We must elect a president who has the concerns about the people, environment, and all living creatures of the USA and also other countries globally. Can we trust this person? Do we believe their agenda? Is this person FOR the people and not just for themselves?

Can this president change and AWAKEN the world?

CHAPTER 7

What Makes a Great President?

By Sean Stewart Price

Source: www.scholastic.com

> "You have probably heard the old saying that "anybody can grow up to be President." But, not everybody is cut out to be President. It takes a special kind of person, someone tough, smart, and driven, just to run for the job. It takes still more talent and character to hold up under the pressures of life in the White House."

And then the author is saying how we expect a lot from a President ... Especially when he/she needs to take quick action on problems facing the nation, such as crime and drug abuse.

> "However, the U.S. Constitution limits the President's power to act. Only Congress can pass legislation, and Congress sometimes moves slowly."

The President can only approve or veto (reject) legislation that Congress passes. Even then, Congress can override a veto and make it the law. The Supreme Court can also limit the President's power by ruling that a law or action violates the U.S. Constitution. "The President has less power than the average voter thinks he does," says presidential expert Paul Boller. "He can't simply by himself make major domestic policies."

The author mentions:

- the ability to face crises and
- "the commonly accepted things that people look for, such as integrity, strength, and caring. "

How important is character in deciding which candidate to vote for? Some experts say that voters today worry more about the issues: what the candidates plan to do about crime, health care, education, and other problems. So candidates, until election day, will be talking about their plans on how they will deal with those major problems. Listen carefully to their agenda.

Consider a leader who will be a champion for the environment and healing of the Earth and all its creatures. Remember corporate greed and the damage to the water, soil, and air which comes from pollution. Vote with your mind and your heart.

Summary

We human beings are 200,000 years old. We have been given a wonderful Gift: our beautiful planet Earth to live in and to be fed by, to explore, and to enjoy ... for us and for our children and future generations, too. But we are also given the intelligence and the understanding to preserve and nurture this home of ours; it is a living vulnerable thing that needs to be loved and taken care of. But – are we? NO!

The facts and stats don't speak strongly in our favor ... We found more ways to destroy it than to sustain it. This book is informing you what we have done to the planet, what we are doing to it currently, and what needs to be done to reverse the damage and to preserve it.

You read things that shocked you and made you MORE aware of what we have done to this beautiful planet. You read many quotes and thoughts from famous people and organizations around the globe, giving their thoughts and opinions about the situation, but also DOING great things to help the situation: to clean and preserve it and inviting everyone else that wants to help to join them!

Do your own daily earth-saving tasks, and set your intention that YOU ARE saving your world, "our common home", get excited about it and always REMEMBER: you *are never too small* to make a difference in our world!

And – don't worry about *the others*: we are like the waves in the ocean or a radio-stations – all energy – people around you *WILL feel* your excitement and your great energy!

MOST importantly, it is my belief and sincere hope that ***this book WILL AWAKEN YOU and inspire you to go and do*** *what we all need to do,* **and to always remember and remind each other to do it**

with love, respect for each other and our differences and with compassion for all living things.

So – let's change the world and do it with love!

Love and many Blessings,

Forever hopeful.

Ilka Handshaw

Appendix

"The mind is as a parachute – it works at it's best when it is OPEN..."

Are you open-minded? I assume you are. Most of us opened-minded people are AWAKENED already, so we need to help others do the same.

Now it is time to read/hear HOW else our thoughts and emotions impact our Earth. If you remember studying physics , especially quantum physics, everything, including you and I, vibrate with a particular energy/frequency. Whatever we think about or carry inside – we vibrate and project it all around us and *other* people or animals and plants , that can ***feel it and are affected by it.***

For example: negative emotions (fear, anger, jealousy) will lower your body's frequencies by 12 MHz. When you think positively again, you'll rise another 10 MHz, and when you pray, you will raise your energy by 15 MHz! (see research on this on my site: *www.ilkashealthyheaven.com*)

So – let me support this by a bit of a science! You have most likely heard about the polygraph aka lie detector, right? The man behind it is Cleve Backster one of the first to do methodical research on cell structure responding to thought. "Cleve B. created the world standard in all polygraph work known as the Backster Zone Comparison. When a human is attached to a polygraph, or a lie detector, it picks up the most minute electrical impulses." * (From the book *"Your Empowering Spirit"* by Gary Sinclair, Cyberphysiologist; p.45.)

At a cellular level there is an increase in the energy fields when a person processes any information and "there is difference in the field when you tell something that you know to be absolutely true and something that is untrue". And Cleve Backster is the unquestioned leading authority in understanding all of this.

He also *worked on plants!* At his first *accidental* plant energy findings on a Darcena plant in 1966, he attached electrodes to the plant leaf to determine how long it takes water to go from the base of the root to the tip of the leaves. He got answers that he never expected (MORE then what he expected). The more he worked with the plant and followed the polygraph charts, *the more "human" the plant's reactions seemed to be.*

"He started to think about burning the plant with a match. At the very instant of that burning creative thought, the recordings on the chart showed a response that again was indicative *OF HUMAN MENTAL RESPONSE...* ***Could it be that plants can read our mind? Could this mean that the simplest living cells are all able to read our thoughts?***

Apparently so, as the research went on to show." Yes, nature reacts to our thoughts! Another example that Gary Sinclair explained to us (He came to Novi, MI years ago to teach us, a group of lifetime learners, all of this that I'm sharing now. An example is how we can CALM Mother Nature with our thoughts (and I can add – prayers). He talked to us about a case when he was teaching a class on ***our ability to use the power of OUR THOUGHTS to transfer healing energy*** as a gift for healing ourselves and others.

When a weather phenomena was taking place, and Hurricane Fran was about to pass by. people were told by local reports that its speed was increasing from 115 MPH and was expected to reach 120 MPH in the next 20 minutes as the eye was making land further down land. He directed all those that were there to ***face the eye of the hurricane and with the mind's eye, to tell the winds to go in the reverse direction***. Twenty minutes later, when the eye hit land, the winds had calmed down to 85 MPH. (*Association For Research and Enlightenment, Venture Magazine, May/June 1997)*

Of course, the lives of those that night were changed, experiencing and *recognizing the Power we have in group consciousness by thought to change things in our lives, including weather (hurricanes and other disasters). Visualize what you want to happen! Use your imagination!*

"Imagination is more powerful than knowledge"

—ALBERT EINSTEIN.

Another great proof of how what we think affects everything around us is Dr. Masaru Emoto's research in his book "*The Hidden Messages in the Water*".

These are all examples of how we can AFFECT Mother Nature in either way – good or bad. But it would be great if we ONLY affect her in a GOOD way…!!! So – use you power in a great way – Mother Earth gives us everything we need; it just make sense to love her!

Until the next time – seeing you at a class or – until my next book…

For now – let's get busy. We have a beautiful planet to save … and savor!

ABOUT THE AUTHOR

Ilka Handshaw was born in Bulgaria, studied languages in high school and graduated from Veliko Turnovo University in Bulgaria. She loved being a teacher, but also loved helping people to be healthier and happier, and always learning on this subject since a child.

She always had an amazing affinity for Mother Nature, absolutely adored it, and since she was a young child, has spent as much time being within Mother Nature by hiking and being on the shores and sunny beaches of the beautiful Black Sea.

Here, in USA, she continued working with her languages as a tutor and interpreter, but is also in the process of learning more on the subject of personal healing. Ilka has started helping others learn to feel better and teaching them how to be more joyful by fulfilling their purpose in life.

She is an author of many articles on health and her passion of essential oils in Michigan magazines. She began writing a book on the subject of health but turned in the direction of inspiring others to save our planet Earth from destruction which humans have caused.

And that in many ways is aligned with personal healing, because as we heal ourselves, so do we heal the world!

Ilka Handshaw
616 259-7509

Resources of Related Interest/ Recommended Reading

On AWAKENING:

"Soul Shifts", Barbara De Angelis; best-selling author of 14 books, including *"How did I get Here?" and "Real Moments"*

"Spiritual Selling", Joe Nunziata

"The System for Soul Memory", Susan Kerr

"What to Say When You Talk to Yourself", Dr. Shad Helmstetter, Ph. D.

"Your Empowering Spirit", Gary Sinclair, Cyberphysiologist

"The Secret life of Your Cells", Dr. Robert Stone, Ph. D.

"The Hidden Messages in the Water", Dr. Masaru Emoto.

On facts/stats of our planet:

www.NASA.com;
www.NOAA.gov
www.foodandwaterwatch.com
www.ipcc.ch (Intergovernmental Panel on Climate Change)
www.worldometers.info
www.Greenpeace.org

For Emotional release/Ancestral Clearing classes or to invite Ilka to speak on HOW TO HAVE A HEALTHY Planet – contact her by e-mail or phone.

You can also find some info on classes on her "healthy" website: www.ilkashealthyheaven.com

Please, feel free to give feedback, share thoughts, ask questions!
616 259-7509; info@ilkashealthyheaven.com

74695239R00055

Made in the USA
Middletown, DE
29 May 2018